Capillary Fluid Exchange

Regulation, Functions, and Pathology

Colloquium Lectures on Integrated Systems Physiology-From Molecules to Function

Editors

D. Neil Granger, *Louisiana State University Health Sciences Center*
Joey P. Granger, *University of Mississippi Medical Center*

Physiology is a scientific discipline devoted to understanding the functions of the body. It addresses function at multiple levels, including molecular, cellular, organ, and system. An appreciation of the processes that occur at each level is necessary to understand function in health and the dysfunction associated with disease. Homeostasis and integration are fundamental principles of physiology that account for the relative constancy of organ processes and bodily function even in the face of substantial environmental changes. This constancy results from integrative, cooperative interactions of chemical and electrical signaling processes within and between cells, organs and systems. This eBook series on the broad field of physiology covers the major organ systems from an integrative perspective that addresses the molecular and cellular processes that contribute to homeostasis. Material on pathophysiology is also included throughout the eBooks. The state-of the art treatises were produced by leading experts in the field of physiology. Each eBook includes stand-alone information and is intended to be of value to students, scientists, and clinicians in the biomedical sciences. Since physiological concepts are an ever-changing work-in-progress, each contributor will have the opportunity to make periodic updates of the covered material.

Published titles
(for future titles please see the website, www.morganclaypool.com/page/lifesci)

Capillary Fluid Exchange: Regulation, Functions, and Pathology
Joshua Scallan, Virgina H. Huxley, and Ronald J. Korthuis
2010

The Cerebral Circulation
Marilyn J. Cipolla
2009

Hepatic Circulation
W. Wayne Lautt
2009

Capillary Fluid Exchange: Regulation, Functions, and Pathology

Joshua Scallan, Virgina H. Huxley, and Ronald J. Korthuis

www.morganclaypool.com

ISBN: 9781615040667   paperback
ISBN: 9781615040674   ebook

DOI 10.4199/C00006ED1V01Y201002ISP003

A Publication in the Morgan & Claypool Life Sciences series Colloquium lectures
*SYNTHESIS LECTURES ON INTEGRATED SYSTEMS PHYSIOLOGY—
FROM MOLECULES TO FUNCTION*

Lecture #3
Series Editors: D. Neil Granger, *Louisiana State University Health Sciences Center*
                Joey P. Granger, *University of Mississippi Medical Center*
Series ISSN
Colloquium Lectures on Integrated Systems Physiology-
From Molecules to Function
Print 1947-945X   Electronic 1947-9468

Capillary Fluid Exchange

Regulation, Functions, and Pathology

Joshua Scallan, Virgina H. Huxley, and Ronald J. Korthuis
University of Missouri-Columbia

*COLLOQUIUM LECTURES ON INTEGRATED SYSTEMS PHYSIOLOGY—
FROM MOLECULES TO FUNCTION #3*

 MORGAN & CLAYPOOL LIFE SCIENCES PUBLISHERS

ABSTRACT

The partition of fluid between the vascular and interstitial compartments is regulated by forces (hydrostatic and oncotic) operating across the microvascular walls and the surface areas of permeable structures comprising the endothelial barrier to fluid and solute exchange, as well as within the extracellular matrix and lymphatics. In addition to its role in the regulation of vascular volume, transcapillary fluid filtration also allows for continuous turnover of water bathing tissue cells, providing the medium for diffusional flux of oxygen and nutrients required for cellular metabolism and removal of metabolic byproducts. Transendothelial volume flow has also been shown to influence vascular smooth muscle tone in arterioles, hydraulic conductivity in capillaries, and neutrophil transmigration across postcapillary venules, while the flow of this filtrate through the interstitial spaces functions to modify the activities of parenchymal, resident tissue, and metastasizing tumor cells. Likewise, the flow of lymph, which is driven by capillary filtration, is important for the transport of immune and tumor cells, antigen delivery to lymph nodes, and for return of filtered fluid and extravasated proteins to the blood. Given this background, the aims of this treatise are to summarize our current understanding of the factors involved in the regulation of transcapillary fluid movement, how fluid movements across the endothelial barrier and through the interstitium and lymphatic vessels influence cell function and behavior, and the pathophysiology of edema formation.

KEYWORDS

Starling equation, capillary filtration, permeability, microvascular network, extracellular matrix, intersitital flow, lymphatics, edema, edema safety factors

Contents

CHAPTER 1

Fluid Movement Across the Endothelial Barrier

The partition of fluid between the vascular and interstitial compartments is regulated by forces (hydrostatic and oncotic) operating across the microvascular walls and the surface areas of permeable structures comprising the endothelial barrier to fluid and solute exchange, as well as within the extracellular matrix and lymphatics. Under normal conditions, the balance of forces acting across the microvascular walls of exchange vessels favors the net flux of fluid from the bloodstream to the interstitium, a process commonly referred to as capillary filtration. (In this treatise, the term capillary filtration is used in an operational sense to refer to volume flow into the interstitium across all vessels (capillaries and postcapillary venules) involved in this transfer of fluid.) The transcapillary flow of fluid contributes to volume flow into lymphatic vessels draining tissues. Because plasma proteins also cross the microvascular barrier to gain access to the interstitial spaces, the flow of lymph provides the only route for return of these extravasated macromolecules to the blood circulation. Were it not for this outflow pathway, the accumulation of proteins into the interstitial spaces would soon disrupt the balance of forces regulating transcapillary fluid filtration, resulting in a redistribution of volume and plasma proteins from the vascular space into the interstitial compartment that is incompatible with life.

Alterations in the forces acting across the endothelial cells of exchange vessels and/or the surface area and permeability characteristics of this barrier to movement of fluid and solutes, allows for moment-to-moment regulation of transcapillary fluid flow and thus vascular volume. For example, redistribution of fluid across the microvascular endothelial barrier constitutes a mechanism for removal of excess fluid from the bloodstream in the case of vascular volume overload. In situations where intravascular volume is reduced, such as following hemorrhage, the forces acting across exchange vessels readjust to favor the reabsorption of interstitial fluid into the vascular compartment as an autotransfusion response, thereby aiding in the restoration of blood pressure towards normal levels. In conditions such as inflammation and capillary hypertension, the forces and membrane parameters governing transendothelial flux are altered to favor enhanced filtration such that excess interstitial fluid accumulates, leading to edema formation. Edema can have both positive and negative consequences with regard to tissue function. For example, enhanced transcapillary fluid filtration increases the convective flux of macromolecules (bulk flow carrying macromolecules along with it) such as antibodies or activated complement proteins into the interstitium that aid in killing bacteria following their introduction into the tissue spaces. In addition, increased interstitial fluid volume that forms in response to enhanced capillary filtration during an inflammatory response may dilute cyto-toxic chemicals released by either invading microorganisms, ruptured tissue cells, or leukocytes that emigrate into the tissue spaces. On the other hand, an expanded interstitial fluid volume increases the diffusion distance for oxygen and other nutrients, which may compromise cellular metabolism in the edematous tissue. Interstitial (or cellular) edema can also impair nutritive tissue perfusion by collapsing capillaries in the swollen tissue (capillary no-reflow), especially in encapsulated organs. Edema formation and capillary no-reflow both act to limit the diffusional removal of potentially toxic byproducts of cellular metabolism.

In addition to its role in the regulation of vascular volume, transcapillary fluid filtration also allows for continuous turnover of water bathing tissue cells, providing the medium for diffusional flux of oxygen and nutrients required for cellular metabolism and removal of metabolic byproducts. Transendothelial volume flow has also been shown to influence vascular smooth muscle tone in arterioles, hydraulic conductivity in capillaries, and neutrophil transmigration across postcapillary venules while the flow of this filtrate through the interstitial spaces functions to modify the activities of parenchymal, resident tissue, and metastasizing tumor cells. Likewise, the flow of lymph, which is driven by capillary filtration, is important for the transport of immune cells, tumor metastasis, and return of filtered fluid and extravasated proteins to the blood. Given this background, the aims of this book are to summarize our current understanding of the factors involved in the regulation of transcapillary fluid movement, how fluid movements across the endothelial barrier and through the interstitium and lymphatic vessels influence cell function and behavior, and the pathophysiology of edema formation.

1.1   HYDROSTATIC AND ONCOTIC PRESSURES, SURFACE AREA, AND PERMEABILITY DETERMINE EXCHANGE: THE STARLING EQUATION

In 1896 Starling proposed that fluid movement between different vascular and tissue compartments would cease when the forces governing fluid movement across the barrier are in balance [246]. Despite Landis' [151, 152] verification of Starling's hypothesis nearly 80 years ago, controversy remains as to the sites of fluid and solute movement, the physical, cellular, and tissue mechanisms responsible for solute selectivity, and the extent to which tissue and vascular composition are regulated.

According to the classic view of Starling, passive factors determine exchange. These factors include surface area (S), and gradients in hydrostatic (ΔP) and oncotic ($\Delta \pi$) pressure (Figure 1.1). In the case of a capillary, for example, ΔP represents the pressure in the capillary (P_c) relative to that in the interstitium (P_t),

$$\Delta P = P_c - P_t \, , \tag{1.1}$$

whereas $\Delta \pi$ represent the oncotic pressure difference between the capillary and interstitial compartments,

$$\Delta \pi = \pi_c - \pi_t \, . \tag{1.2}$$

Capillary hydrostatic and interstitial osmotic pressure favors the movement of fluid from the vascular to extravascular compartment, while plasma osmotic pressure represents a force that tends to pull fluid into the vascular space (Figure 1.1). Interstitial fluid pressure in many tissues is slightly negative relative to atmospheric pressure under normal conditions and thus tends to pull fluid into this space. With increasing tissue hydration, however, interstitial fluid pressure becomes positive, which favors fluid movement into the vascular compartment (Figure 1.1). Since fluid movement from the blood to interstitium occurs across capillaries and postcapillary venules, the terms capillary and transcapillary in the above discussion are often used in an operational sense to signify transmicrovascular forces and flows occurring across these exchange vessels.

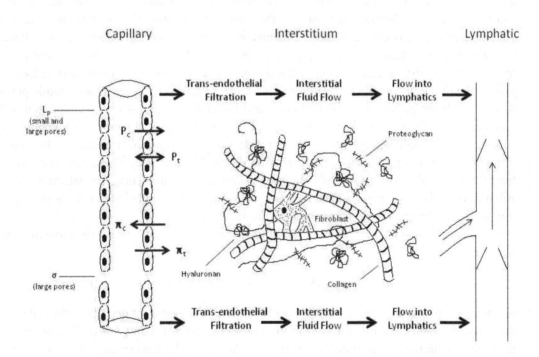

Figure 1.1: Capillary—interstitium—lymphatic fluid and protein exchange system depicting the Starling forces acting across the microvascular wall, the organization of the interstitial matrix, and valvular structures of larger lymphatic vessels. Net filtration pressure, which represents the net sum of the Starling forces, is slightly positive for most organs and results in fluid filtration that is balanced by removal via lymph flow. The Starling forces are the capillary (P_c) and tissue (P_t) hydrostatic pressures and the plasma (π_c) and interstitial (π_t) colloid osmotic pressures. Two membrane parameters, the hydraulic conductivity (L_p) and the osmotic reflection coefficient (σ), and the surface area available for exchange also influence the flux of fluid and solute across the microvascular walls. Modified from references [88] and [279].

Surface area, S (cm^2), an architectural and structural parameter, is determined by the dimensions (diameter and length) and number of vessels in a network. Since only a portion of capillaries are open to flow at any given moment and the number of perfused capillaries is determined by tone in terminal arterioles (which function as precapillary sphincters), capillary surface area available for exchange is subject to active regulation. Similarly, because hydrostatic pressure in the vascular space is a function of blood pressure, arteriolar vascular smooth muscle tone upstream in a microvessel network is the primary moment-to-moment regulator of ΔP. The oncotic pressure π is a direct function of protein concentration in a compartment, as can be seen from Onsager's law:

$$\pi = \Phi n \mathbf{C} \tag{1.3}$$

where Φ is a correction coefficient, n is valence number, and $\mathbf{C}$ is molar concentration and is presumed constant under physiological conditions. In the blood, plasma protein concentration is the major determinant of plasma colloid osmotic pressure, with albumin playing a dominant role. Because albumin is smaller than most other plasma proteins, its relative concentration is comparatively higher in the interstitial fluid, owing to the sieving characteristics of the microvascular barrier to protein flux. Thus, extravasated albumin exerts a greater impact on tissue colloid osmotic pressure than other plasma proteins relative to its effect on plasma oncotic pressure. In addition to its effect on colloid osmotic pressures in the vascular and interstitial spaces, albumin also functions as an antioxidant and facilitates the transport of a wide variety of substances, owing to the presence of many surface-charged groups and ionic/hydrophobic binding sites. These substances include free fatty acids, calcium, phospholipids, bilirubin, enzymes, hormones, drugs, metabolites and ions. Albumin not only serves as a carrier to transport amino acids to the tissues but can also be pinocytosed into cells, providing amino acids to cells following its degradation.

The expression below describing fluid flux (J_v, volume flux, ml s^{-1}) is the modern form of the Starling expression:

$$J_v = L_p S(\Delta P - \sigma \Delta \pi) = L_p S(NFP) , \tag{1.4}$$

where L_p is the hydraulic conductivity (cm s^{-1} mmHg^{-1}, a coefficient describing the leakiness of the barrier to water), NFP is the net filtration pressure, and σ is a unit-less coefficient known as the osmotic reflection coefficient. NFP represents the sum of the hydrostatic and effective osmotic forces acting across the walls of exchange vessels. Under steady-state control conditions, there is a slight imbalance in the hydrostatic and oncotic forces acting across the walls of exchange vessels such that NFP is positive, resulting in net fluid movement (capillary filtration) into the tissues. This is balanced by removal of interstitial fluid via the lymphatics (Figure 1.1). Thus, under steady state conditions, J_V is equal to lymph flow and interstitial fluid volume is constant. If NFP is negative, this signifies that the balance of forces acting across exchange vessel walls favors net reabsorption of fluid from the interstitial to vascular compartment.

The σ varies between 1 and 0 and indicates the likelihood that a molecule approaching a pore (or any structure that conducts fluid) in a membrane will be reflected back from the pore and retained in the vascular compartment. At one extreme, when $\sigma = 1$, the barrier is a perfect semi-permeable membrane that passes only water and excludes all solute (which is reflected back from the pores 100% of the time). In this state, the gradient in osmotic pressure is fully realized. At the other extreme, when $\sigma = 0$, none of the solute is reflected at the barrier. Under these conditions, no gradient in oncotic pressure can exist and the hydrostatic pressure gradient alone drives fluid movement, which simplifies Equation (1.4) to:

$$J_v = L_p S(\Delta P) \,. \tag{1.5}$$

A graph of volume flux per unit surface area (J_v/S) on hydrostatic pressure, according to Equation (1.4), yields a straight line with slope equal to the hydraulic conductivity (L_p) and pressure (x) axis intercept equal to $\sigma \Delta \pi$ (Figure 1.2).

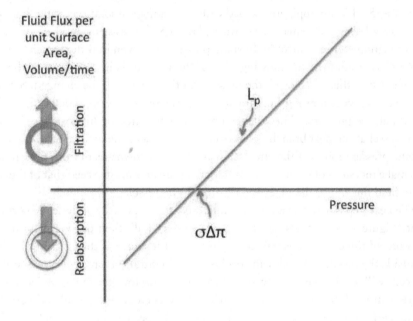

Figure 1.2: Hydrostatic pressure above the effective oncotic pressure gradient results in fluid filtration into the tissue. When the hydrostatic pressure gradient (ΔP) across the vessel barrier equals the effective oncotic pressure gradient ($\sigma \Delta \pi$) no fluid will move (x-axis intercept) and should the pressures fall below that point, fluid will move from the interstitial compartment into the vascular space (reabsorption). The slope of the non-steady state relationship between fluid flux per unit surface area and pressure is the hydraulic conductivity, L_p with units of velocity per unit pressure gradient.

1.2   RELATIONSHIPS BETWEEN TRANSVASCULAR VOLUME FLUX AND HYDROSTATIC PRESSURE: DETERMINATION OF HYDRAULIC CONDUCTIVITY AND EFFECTIVE COLLOID OSMOTIC PRESSURE GRADIENT

The most direct means for assessing hydraulic conductivity, L_p, involves cannulation of a microvessel of known size with a micropipette containing marker particles and measuring volume flux, J_v over a range of hydrostatic pressures. In this case, the constituents of Equation (1.4) are known $(S, \Delta P, \Delta \pi)$, can be approximated reasonably (P_t, π_t), or determined $(L_p$ and $\sigma)$. In practice, this is accomplished by cannulating a microvessel with a pulled glass micropipette with a tip that has been ground to a size slightly smaller than the dimension of the vessel and has a beveled opening – generally 40-55°. The micropipette is filled with a physiological solution containing a protein (usually albumin) and a 1-3% (v/v) hematocrit of washed red blood cells as flow markers. The pipette, attached to a water manometer to control hydrostatic pressure, is driven into the lumen of the vessel so that the vessel slips around the tip allowing flow of fluid and cells into and out of the pipette without obstruction. Once this is achieved, the pressure in the pipette can be changed. Setting the pressure so that the marker cells flow neither into nor out of the tip is a means of determining what is referred to as the "balance pressure." The balance pressure represents the hydrostatic ("blood") pressure of the microvessel at the next branch down from the tip as a cannulated vessel *in vivo* experiences the hydrostatic pressure forces of the circulation at all points downstream from the cannulation pipette. To then make measures of volume flux subsequent pressures must exceed that of the balance pressure to ensure fluid flux out of the pipette and through the vessel.

The next step is to use an additional microtool to temporarily occlude the perfused microvessel segment (Figure 1.3). If fluids can traverse the vessel wall, then on occlusion of the pressurized segment one of three situations will occur. In the first instance, if the pressures in the vessel exceed those outside the vessel, then following occlusion, fluid filtration across the vessel wall occurs and the marker cells will be observed to flow out of the pipette tip towards the occluder. The rate of marker cell movement is a function of the rate at which fluid is crossing the wall and new fluid is traveling out of the perfusion pipette. The closer they are to the occluder, the slower they will move. In the second case, if the pressures inside the vessel are lower than outside, fluid will flow into the vessel and the marker cells will track back into the pipette tip. A third set of conditions occurs when the pressures inside and outside of the cannulated vessel are equal and opposite. Because there is no fluid flux across the vessel wall under this set of circumstances, there is no movement of the marker cells. It is the rate of marker cell movement (dx/dt) that is measured to calculate volume flux, together with estimates of the volume of the vessel relative to its surface area, V/S. For a right cylinder of radius r and length l,

$$V/S = \frac{\pi r^2 l}{2\pi r l} = \frac{r}{2}.$$
(1.6)

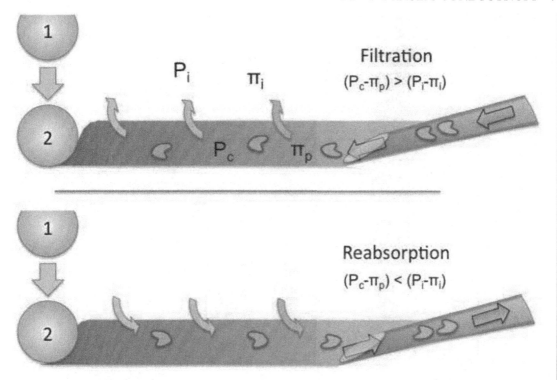

Figure 1.3: Schematic of fluid and marker cell movement during vessel occlusion in the two cases of filtration (top panel) and reabsorption (bottom panel). In the case of filtration, following occlusion of the vessel segment with a glass rod (1 to 2) the net pressure from retained colloids and the hydrostatic pressure in the perfusion pipette exceed the colloid oncotic and tissue pressures resulting in movement of fluid across the barrier. As this fluid leaves the vessel segment new fluid enters the segment from the pipette carrying with it marker red blood cells (or any other suitably sized flow marker). In the lower panel, if the gradient is reversed (more colloids inside the vessel than out, for example), fluid will move from the tissue space into the vessel carrying the marker cells away from the occluding rod, into the perfusion pipette.

Using this information, the volume flux per unit surface area, J_v, is the distance a marker cell moves, Δx, over an interval of time, Δt, having started at a distance, x_o, from the occluding rod in a vessel of radius, r:

$$J_v/S = \frac{1}{x_o} \frac{\Delta x}{\Delta t} \frac{r}{2} . \tag{1.7}$$

These parameters are illustrated in Figure 1.4. When multiple measures of J_v/S are made at several pressures, as stated above, the plot yields a straight line with a slope equal to the hydraulic conductivity, L_p, and x-axis intercept of $\sigma \Delta \pi$, the effective oncotic pressure gradient.

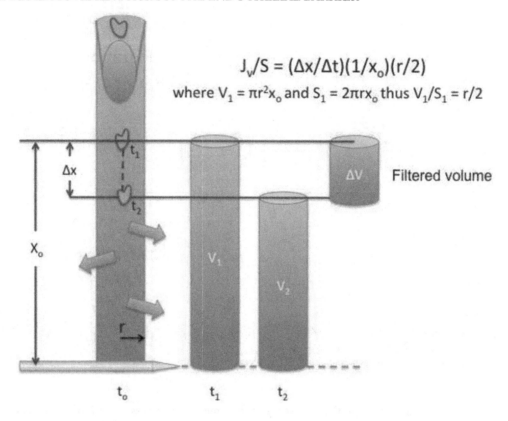

$$J_v/S = (\Delta x/\Delta t)(1/x_o)(r/2)$$

where $V_1 = \pi r^2 x_o$ and $S_1 = 2\pi r x_o$ thus $V_1/S_1 = r/2$

Figure 1.4: The volume filtered from an occluded vessel segment is indicated by the distance a marker cell travels (Δx) in a given amount of time (Δt). The rate of marker cell movement will be a function of distance from the occluding rod – those furthest from the occluder will travel the fastest; those nearest will travel the slowest; within 60 μm of the occluder they may not move at all and the assumption of a circular geometry is likely in error. At the first time point, T1, in a vessel of radius, r, the marker cell is a distance x_o from the occluding rod. The volume of the cylinder at this time is $V_1 = \pi r^2 x_o$. If the vessel wall is permeable to water and a period of time is allowed to elapse, $\Delta t = (t_1 - t_2)$, the marker cell will move a distance Δx. The filtered volume per unit surface area for a cylinder, J_v/S is given as $(\Delta x/\Delta t)(1/x_o)(r/2)$.

This method can be used to determine, in a living vessel, whether a treatment results in a change in hydraulic conductivity. Following a treatment the slope can be seen to increase or decrease with no change in the x-axis intercept, conditions which imply that the barrier has become leakier (increased slope) or tighter (decreased slope) by a mechanism that changes the area available for fluid movement without changing the selectivity or size of the pathways conducting fluid (i.e., σ is a constant). Examples include changes in L_p that occur in response to atrial natriuretic peptide

(ANP), wherein exposure to the hormone results in a rise in L_p and no change in σ [105], and to norepinephrine which caused L_p to fall without any apparent change in σ [110].

Treatment with an inflammatory agent usually results in two changes in the relationship of volume flux and pressure. The slope of the line generated by plotting J_V/S versus pressure increases (i.e., σ hydraulic conductivity is increased) and the x-axis intercept decreases (i.e., σ is reduced). In this case, the structures conducting fluid across the vascular barrier have become larger allowing both fluid and proteins to move across the barrier. Examples of this are more numerous, and include responses to histamine, bradykinin, thrombin, reactive oxygen and nitrogen species, VEGF, TNF-alpha, FMLP, PMA, to mention but a few examples [139, 163, 171, 174, 187]. A similar response is observed upon removal and replacement of the protein albumin (L_p increases on protein removal and σ goes down) [104]. While it is obvious that an inflammatory agent that increases L_p and reduces σ is likely to be detrimental for organ function, because edema formation increases the diffusion distance for oxygen and other nutrients, it is also important to recognize that the opposite, a reduction in L_p, can also result in tissue damage as a consequence of limiting fluid flux into the spaces around metabolizing cells, thereby reducing the removal of metabolic wastes such as hydrogen ion and lactate, among other materials.

1.3   ASSESSMENT OF DIFFUSIVE PERMEABILITY

Another approach is to measure the permeability of microvessels to solutes. Solutes move across the vascular barrier when a gradient in concentration exists as given by Fick's first law of diffusion:

$$J_s = P_d S(\Delta C) . \tag{1.8}$$

In this case, J_s is the flux of solute (mmole s^{-1}) and P_d is the diffusive permeability coefficient (cm s^{-1}) which is defined as:

$$P_d = \frac{D_f}{\Delta x} , \tag{1.9}$$

where D_f is the free diffusion coefficient for the solute (cm^2 s^{-1}) and Δx is the barrier thickness (cm). In this case, ideally, the solute is a dye or a molecule to which a marker can be attached. Classically, the markers used were radioisotopes. However, more recent work has relied on approaches wherein radioisotopes were replaced by fluorescent dyes. The detection system is then a photometer or digital camera that measures either optical density or fluorescence intensity. Fluorescence techniques have a larger dynamic range, greater sensitivity, and can use light sources in the visible or near infrared and limit photo-damage to living tissue [23, 222, 228]. To measure solute flux (J_s) and calculate permeability (P_d), a vessel is cannulated with a micropipette that either has two chambers (theta pipette) or is fitted with tubing to control the composition of the fluid flowing in the vessel [98] (Figure 1.5). By this approach, the contents of the pipette are under pressure that can control which fluid perfuses the vessel lumen as well as controlling the hydrostatic pressure in the vessel segment. Again, when pressure in the pipette is adjusted so that the dye front is neither moving into or out of the pipette tip, one has a measure of the "balance" pressure and the pressure at the next bifurcation

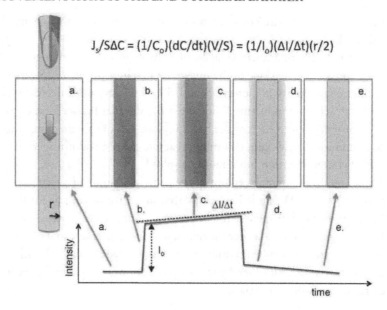

$$J_s/S\Delta C = (1/C_o)(dC/dt)(V/S) = (1/I_o)(\Delta I/\Delta t)(r/2)$$

Figure 1.5: Movement of solutes from a perfused vessel segment can be visualized using optically dense or fluorescent dyes. Quantitative measurement of solute flux (J_s) requires perfusion of a vessel segment of known radius (r) with a constant and known concentration of labeled solute (C_o) and the ability to measure optical density (OD) or fluorescence intensity (I). The process of perfusing a vessel segment to quantify J_s is illustrated in panels A-E corresponding to the tracing of Intensity against time below. In panel A the vessel segment is perfused with a non-dye containing physiological salt and protein (perfusion) solution from one half of a theta pipette. No labeled material flows from the second half of the pipette (containing the same as the perfusion side with the addition of dye-labeled solute: dye) during this time because the pressures are set to achieve "balance" inside the vessel lumen with the contents of the pipette. During this time a baseline reading of intensity is achieved (A) over an area of interest (AOI) containing a length of vessel and surrounding tissue. In (B) the pressures in the two halves of the theta pipette are reversed (dye high, perfusion balanced) resulting in a step increase in fluorescence intensity (I_o). If the solute in question can leave the vessel space and move into the tissue (C) there will be a further increase in I representing the material flowing through the vessel plus that in the tissue such that the rate of change in I is $\Delta I/\Delta t$. The process is reversible; switching pressure back to that in (A) there is a drop in I that should be as large as I_o but offset from the original baseline by the amount of material in the interstitium. With time, if the labeled material is free to diffuse, I should return to baseline (E). The baseline I will be elevated If the solute were to be retained in the tissue due to binding or sequestration. Control experiments need to be performed to ensure that the excitation light intensity does not oxidize the fluor during the measurement (resulting in fade and generation of reactive oxygen species), that the dye concentrations are neither too high (internal quenching) or too low such that concentration is no longer a linear function of fluorescence intensity, and that the vessel is circular in cross section (otherwise V/S is not $r/2$).

of the network. Obviously to make a measure of flux it is necessary to raise the pressure to ensure complete, rapid filling of the vessel segment (Figure 1.4(b)). Failure to fill or empty the vessel segment results in the inability to estimate the intravascular concentration that is required to calculate flux. The germane elements are vessel radius (r), the step change in intensity (I) from perfusion with solution devoid of labeled solute to perfusion with the fluor-labeled solute (I_o, Figure 1.5) and the slope of the tracing as dye flows through the vessel segment and moves into the surrounding tissue ($\Delta I / \Delta t$). Accordingly,

$$J_s / S \Delta C = P_s = \frac{1}{I_o} \frac{\Delta I}{\Delta t} \frac{r}{2} \, , \tag{1.10}$$

where again the assumption is that the vessel segment is circular in cross section so that $V/S = r/2$ as in Equation (1.6). The area over which fluorescence intensity is monitored is ideally over a straight length of vessel of uniform radius and over a tissue area wide enough that dye labeled solute does not leave the window. Additional calibrations, considerations and limitations of the approach are discussed in a number of publications [7, 47, 83, 106, 108, 137, 230, 231]. In addition, it is important to choose a fluorophore that does not excite at a wavelength that is absorbed by other biological materials (fluorescein or the Alexa 488 dyes emit in the region where the iron in the heme on hemoglobin and myoglobin absorb, for example). It is also important that the dye not fade during the time of measurement (leading to an underestimate of the flux and the potential for the production of oxidants that interfere with normal function). The techniques described above have been optimized for use in the setting of confocal microscopy wherein the mathematics for the geometry of a vessel in a confocal slice are more complicated but eminently approachable [230].

1.4 CONVECTIVE SOLUTE TRANSPORT ACROSS THE MICROVASCULAR WALLS

In the ideal case of Fick's First law (Equation (1.8)), the only mechanism moving solutes across a barrier is free diffusion (Equation (1.9)); where the direction of movement is governed by a concentration gradient (from higher to lesser amount). This is then considered to be a "passive process" and a plot of J_s / S as a function of pressure is a straight line with zero slope and an intercept on the y-axis that is equal to P_d (Figure 1.6).

In actual fact, both water and solutes move across vessel walls at the same time; in some cases, through the same pathways (Figure 1.1). As a consequence, in the intact circulation, both hydrostatic and oncotic pressure gradients are functions of solute and water movement [45, 174]. When this is the case, σ is not equal to 1, and the movement of fluid can influence the movement of solute by a process known as "solvent drag," solute coupling, or convective coupling. Under such conditions the plot of J_s / S on ΔP is non-linear (Figure 1.6) and described by the following equation:

$$J_s / S \Delta C = P_d \left[\frac{Pe'}{(e^{Pe'} - 1)} \right] + J_v / S (1 - \sigma) \, , \tag{1.11}$$

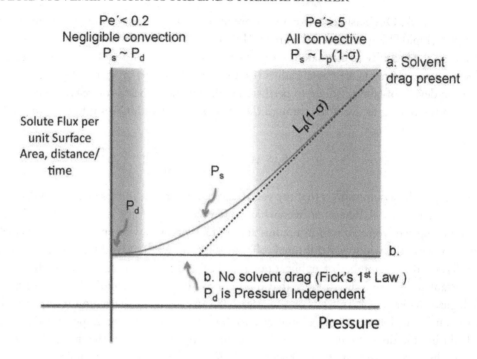

Figure 1.6: If solute flux increases with increasing pressure water and solute are travelling together. The assumption is that solute flux (J_s) only occurs when there is a concentration gradient (ΔC) across a barrier with permeability, P_s. If solute and fluid travel by the same pathways and the diffusional rate of solute movement is of a similar or lesser magnitude that the rate of volume flux, as pressures increase fluid can carry solute (convective coupling or solvent drag). When this occurs a plot of J_s against pressure will not have a zero slope (Fick's first law) but instead be curvilinear with a limiting slope equal to the hydraulic conductivity and reflection coefficient of the pathway conducting water and solute. The Pe' number is the unitless ratio of volume flux relative to diffusive flux. When $Pe' < 0.2$ diffusive forces predominate and the flux appears pressure-independent (shaded area to the left); at $Pe' > 5$ the convective forces prevail and solute flux is a linear function of pressure.

where Pe' is the $Pe'clet$ number:

$$Pe' = \frac{\left[\dfrac{J_v(1-\sigma)}{S}\right]}{P_d}. \tag{1.12}$$

Assuming a homoporous barrier and a small constant interstitial oncotic pressure (first approximations),

$$P_s = \frac{J_s}{S\Delta C} = P_d \left[\frac{Pe'}{(e^{Pe'} - 1)} \right] + L_p(1 - \sigma)(\Delta P - \sigma \Delta \pi) . \tag{1.13}$$

The $Pe'clet$ number provides a means for determining how much of the net flux of solute is governed by fluid movement relative to solute movement. When Pe' is greater than 5 (i.e., volume flux is at least three times the diffusive flux), the primary means by which that solute moves is by "convection" or "solvent drag" – i.e., with water flow and Equation (1.11) reduces to:

$$J_s/S\Delta C = L_p(1 - \sigma)(\Delta P - \sigma \Delta \pi) . \tag{1.14}$$

In this condition, the limiting slope is $L_p(1 - \sigma)$, which provides a measure of the water conductivity and reflection coefficient of the pathway that conducts water and the solute in question (Figure 1.7).

At the other extreme, when $Pe' << 1$ (generally $Pe' < 0.2$), the measures of flux provide a value for permeability not different from P_d (Figure 1.7). Caution is required when measuring flux and calculating P_s without knowledge of the pressures because when σ for the solute is not equal to or close to 1, convective coupling is appreciable and can be the dominant mechanism behind perceived changes in exchange in the absence of a real alteration in barrier function. This happens most often when flux is measured in an intact circulation without the ability to control the number of the vessels perfused or blood pressure and is the basis for suspicion regarding the interpretation of data obtained using the Miles assay or by assessment of transcapillary escape rates for labeled proteins (in these approaches, a dye which binds to albumin or radiotracer-labeled proteins are introduced into the circulation, an agent is applied, and then an ear, a segment of skin, or other thin tissue is observed for accumulation of dye, or biopsies are obtained for radioactive counting, respectively). When convection occurs, more material will egress the circulation if blood pressure increases in response to the agent in question, if the agent in question dilates smooth muscle and increases perfusion into more of the vasculature, and/or if the agent increases P_d or reduces σ.

With respect to volume flux, if the pressures in the microvessels keep changing so that there is insufficient time for solute to move across the barrier and change the gradients across the barrier, the modified Starling equation (Equation (1.5)) holds and the plots are similar to that depicted in Figure 1.2 or Figure 1.6, with a slope equal to the hydraulic conductivity and the pressure axis intercept equal to $\sigma \Delta \pi$. If instead solute and fluid can flux across the walls, the gradients will change and a steady state condition can be achieved wherein the transvascular osmotic gradient influences steady-state filtration as given by

$$J_V = L_p S \left(\Delta P - \left(\sigma^2 \pi_p \left[1 - e^{-Pe} \right] \right) \left(1 - \sigma e^{-Pe} \right)^{-1} \right) . \tag{1.15}$$

The steady state filtration relationship is non-linear. At high pressures ($Pe' > 5$), the relationship of J_v/S is linear and parallels the transient, Starling relationship with a limiting slope equal

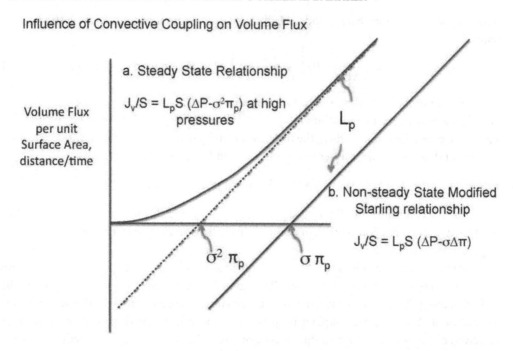

Influence of Convective Coupling on Volume Flux

Figure 1.7: Not only can fluid flow influence solute flux but solute flux can influence fluid movement. Starling's Law of Filtration, $J_v = L_p S(\Delta P - \sigma \Delta \pi)$, is a non-steady state relationship (b) assuming that the forces in the vessel and the tissue remain unchanged by the movement of fluid and/or solute. A graph of Jv/S against pressure is linear having a slope of L_p and a pressure axis intercept of $\sigma \Delta \pi$. In the intact tissue, as water moves across the barrier it can dilute osmotically active solute thus changing the oncotic pressure. Similarly, if the colloids responsible for the oncotic pressure permeate the barrier and follow their gradient, fluid will follow also resulting in changes in the pressure distributions. In vessels where the pressures are relatively constant (as occurs in the exchange microcirculation) a steady state will be achieved giving rise to the solid curvilinear relationship. In the steady state, reabsorption cannot be maintained and vessels either have no fluid movement across the wall or filter fluid. Only transient changes in hydrostatic pressure will produce reabsorption when pressures drop below $\sigma^2 \Delta \pi_p$. In either case, the limiting slope is the hydraulic conductivity of all pathways conducting fluid.

to the hydraulic conductivity (Figure 1.7). If J_v is measured at a pressure where $Pe' > 5$ and the hydrostatic pressure is reduced rapidly to a value below the pressure axis intercept, $\sigma^2 \pi_p$, transient reabsorption can be observed with marker cells moving away from the occluding rod back into the perfusion pipette. The other notable feature of the steady state relationship is that it predicts that vessels, such as capillaries and venules, which are exposed to fairly constant hydrostatic pressures with no noticeable pulsation, will have a negligible to appreciable constant filtration, but no sustained reabsorption of fluid from the interstitium into the vasculature. This is contrary to conventional and widespread perceptions. The one organ for which this will not be the case is the heart as pressures in the microvasculature rise and fall with each beat [170]. Other widely held perceptions of exchange are that: 1) arterioles do not participate in exchange; 2) capillaries, given their large number, are the primary site for volume flux (J_v); and 3) venules, given their "leaky" walls and low pressure, are the major site for protein flux (J_s). Despite extensive evidence to the contrary that has accumulated over the past two decades, these concepts are still widely espoused in textbooks.

1.5   CAPILLARY THE MORPHOLOGY VARIES AMONG ORGANS IN ACCORD WITH FUNCTION

Because of the perception that capillaries and venules are considered to be the primary sites for fluid and solute exchange, we will review the morphology of these vessels. While the capillaries are the smallest elements of the vasculature in terms of diameter, they are also the most numerous. As a consequence, capillaries have the highest cumulative surface area available for exchange. Given their small size (for mammals the diameter is on the order of 3 to 6 μm) and the Windkessel function of upstream arterial vessels, blood flow is generally non-pulsatile in capillaries and the erythrocytes flow through these minute vessels one-by-one. Exchange of oxygen occurs primarily from the red cell "packets" as they pass through this portion of the vasculature. On the other hand, from the plasma gaps between the red cells pass carbon dioxide, fluids and molecules up to the size of the plasma proteins.

The endothelial lining of all segments of the vasculature appears to possess a luminal electron lucent layer referred to as the glycocalyx (or fiber matrix) that can act as a physical barrier at the blood/endothelial interface. The glycocalyx was first observed by Luft [162] over 40 years ago and has become the subject of intense current research interest because of its role as a barrier to transvascular solute flux, ability to influence leukocyte rolling, stationary adhesion and emigration, function as a mechanosensor to regulate vascular tone, and as a depository for growth factors and enzymes. Structurally, the glycocalyx consists of a negatively charged (anionic) three-dimensional mesh-like layer of glycoproteins and glycolipids that are arranged in a quasi-periodic repeating substructure in both the radial and axial directions [15, 36, 43, 45, 59, 93, 195, 216, 242, 259]. The components of the glycocalyx have been implicated in cytoskeletal organization as both syndecan, the primary heparin sulfate proteoglycan, and podocalyxin, the primary sialoprotein, modulate cell-cell and cell-matrix adhesion through their cytoplasmic domains [68]. Changes in hemodynamic environment as well as the presence and absence of vasoactive mediators appear to alter the depth, distribution, and three-dimensional architecture of the glycocalyx resulting in changes in the flux of materials across the barrier [107, 126, 179, 195]. Model analysis based on the assumption that the glycocalyx acts as the principal molecular sieve for plasma proteins [6] may aid in accounting for the disparity in measured lymph flows versus calculated values derived from careful measurements of tissue osmotic and hydrostatic pressures using sophisticated new methods, as outlined by Levick [159]. Levick's provocative analysis suggested that no reabsorption occurs on the venous side of capillaries for most vascular beds (notable exceptions were the kidney and intestinal mucosa, organs which must reabsorb fluid to support their function), making it difficult to reconcile with low in vivo values for whole-body lymph flow rates. Based on Levick's observations, Michel [175] and Weinbaum [102] proposed that the glycocalyx represents the effective osmotic barrier, not the entire wall of exchange vessels, as has been previously assumed. This forced a reconsideration of the original Starling hypothesis, wherein it is proposed that the oncotic Starling forces are determined by the local difference in protein concentration across the glycocalyx alone and not the global difference in luminal and tissue colloid osmotic pressures [102, 158, 275, 289]. Subsequent work suggested that small extravascular compartments near the vessel wall (below the glycocalyx and also trapped by pericytes) transiently regulate transport [289]. More recent studies using indicator dilution dispersion techniques allow assessment of the effects of changes in the glycocalyx on the permeation of small solutes and water flow independent of downstream barriers to these fluxes [76]. This analysis suggests that axial water movement through the glycocalyx exerts little influence on diffusion limited radial transport of small solutes, indicating that the principal determinant of solute dispersion in this layer is the effective diffusion coefficient.

Capillaries, the smallest vessels in the microvascular network, are classified as continuous, fenestrated and discontinuous. In continuous and fenestrated capillaries, the components that influence the amount of material moving are the structures within and between endothelial cells and the position of mural cells (pericytes, fibroblasts, and mast cells) relative to the endothelial cells. These two sets of features are organ dependent and therefore related to the function of the organ. Transendothelial structures include the intercellular junctions, transendothelial "pores," and vesiculo-vacuolar organelles (VVOs), fenestrae, water channels (aquaporins), and vesicles or caveoli [67, 87, 176, 177]. The number and distribution of these structures in the capillary wall allow there to be gradients of "leakiness" within the classifications of continuous and fenestrated vessels.

The continuous capillaries make up the largest class of blood vessels as they populate the two largest organs of the body, skin and skeletal muscle. The tightest continuous capillaries form the blood/brain barrier (BBB), the blood-aqueous barrier, (BAB), the blood-nerve barrier (BNB) and the blood/testes barrier. Of these the blood/testis barrier is the most straight forward and least "malleable." It separates blood from the seminiferous tubules via specialized junctional complexes between adjacent Sertoli cells in the vicinity of the base of the seminiferous epithelium, thereby preventing gametes from circulating in blood in addition to its involvement in spermatogenesis [283].

The BBB and the BAB involve formation of specialized adherens and tight junctional structures between adjacent endothelium. The BBB is a highly specialized structure restricting movement of materials between the vascular and cerebrospinal fluid compartments of the brain and spinal cord. Under normal circumstances, the structures comprising the BBB are very restrictive and few vesicular structures are observed. The existence of the BBB requires the presence of specialized transporters to facilitate one-way and selective movement of required nutrients, particularly glucose, and other small solutes. However, the components that endow the BBB with its highly restrictive characteristics that largely prevent small and large solute extravasation and limit migration of blood-borne cells under normal circumstances are changed in many pathological conditions (stroke, CNS inflammation, and neuropathologies including Alzheimer's disease, Parkinson's disease, epilepsy, multiple sclerosis, brain tumors). In such conditions, BBB disruption ("opening") can lead to increased paracellular permeability and the junctions can be modified to facilitate entry of leukocytes into brain tissue, both of which contribute to edema formation. In parallel, there are changes in the endothelial pinocytotic vesicular system resulting in the uptake and transfer of fluid and macromolecules into brain parenchyma [245, 276]. Cerebral edema is a particularly important pathologic event in the brain because the bony cranial vault prevents expansion of the tissues. As a consequence, edema formation results in a dramatic rise in interstitial pressure, which in turn, acts to physically compress cerebral vessels, thereby compromising blood flow delivery to this highly metabolic organ.

Astrocytes are located in close proximity to the endothelial cells, and as a consequence are also thought to play an important role in the regulation of BBB permeability. In this regard, it is of interest to note that while increases in permeability to small solutes can be observed in model systems of astrocytes co-cultured with brain microvascular endothelium on removal of the astrocytes, endothelial tight junction opening and changes in molecular composition (claudin-3, -5, occludins, ZO-1 or ZO-2 or adherens junctions-associated proteins, beta catenin and p120cas) were not observed to change [94].

The blood-aqueous barrier (BAB) exists in the eye – a semi-permeable membrane consisting of non-pigmented layers of epithelia of the ciliary body joined at the apical surface to endothelia of the iris which serves to restrict movement of materials between the circulation and the aqueous fluid. Some of the pathologies with sight in type II diabetes appear to involve breakdown of this barrier [25, 80]. The tight and adherens junctions identified in the blood-aqueous barrier include occludins, ZO-1, and the cytoskeletal protein, actin [241].

The least studied of "tight" barriers of continuous capillaries is the BNB – involving endothelial cells of peripheral vascular nerves [229]. A current working hypothesis is that the BNB properties are a function of the endothelium in union with peripheral nerve pericytes that expressed peripheral bands of ZO-1 and occludin. In culture, these pericytes demonstrated a relatively high electrical resistance and low clearance of inulin as well expression of a number of barrier-related transporters (ABCG2, p-gp, MRP-1, and Glut-1) that are thought to be involved in the establishment and maintenance of peripheral nerve homeostasis [237]. As with the brain, there is reason to maintain a BNB in the vicinity of the neuromuscular junction to avoid spilling of neurotransmitters into the circulation.

Water may have the ability to traverse the endothelial barrier by yet another set of structures: the aquaporins [31]. Aquaporin-1, a channel that conducts water, has been identified in continuous capillary endothelium in intestine and mesentery as well as in lymph nodes [60, 75]. Aquaporin-4, on the other hand is associated with BBB and CNS endothelia and its expression can be modified by many of the agents demonstrated to result in brain edema [35, 133, 226].

Macromolecules may be able to access the caveolar or vesicular structures in capillary endothelium. While morphological and functional data point to the existence of this pathway [207], there remains a lively discussion as to whether proteins actually use this pathway or the structures forming functional "pores" through and between endothelium [174, 176, 177].

Next in line with respect to solute restriction are the skin, lung, cardiac, skeletal muscle and adipose capillaries. Across these vessels, water and small solutes (such as ions, glucose, amino acids) pass without appreciable restriction (i.e., $\sigma \sim 0$), whereas for the larger proteins, starting with albumin σ is on the order of 0.8 or higher).

Examples of organs, tissues, or structures with fenestated capillaries are the kidney, area postrema, carotid body, endocrine and exocrine pancreas, thyroid, adrenal cortex, pituitary, choroid plexus, small intestinal villi, joint capsules, and epididymal adipose tissue. The fenestrae are structures that generally penetrate the endothelium and form an apparently open, wagon wheel structure (.020 − .100 μm, diameter) and conduct fluid with considerable ease. Kidney fenestrae appear to have two types, one with diaphragms like other organs, the other in the glomerula which appears to be devoid of bridging structures [111]. Of interest, fenestrated capillaries express high levels of both VEGFR-2 and VEGFR-3, and possess normal pericyte coverage [128]. Also, just because capillaries of an organ may not generally possess fenestrae, this does not mean that they cannot express fenestrations. As implicated in the current literature, the endothelium of fenestrated capillaries possess VEGF receptors and continuous capillaries exposed to VEGF can form fenestrae [219]; further a study of skeletal muscle morphology following immobilization demonstrated the appearance of fenestrations in a small, but significant number of the continuous skeletal muscle capillaries [192].

Finally, capillaries in organs involved in the sequestration of formed vascular cells (spleen, bone marrow) or in the synthesis and degradation of fats and proteins (liver) are characterized by wide spacing between endothelial cells (up to microns in diameter). It is interesting to note that an organ such as the liver is vulnerable to "flooding" should venous pressure and/or volume increase, resulting in the formation of ascites fluid. Exchange in the liver is indeed unique given its perfusion by highly oxygenated blood from the hepatic artery and via the nutrient laden, but oxygen-poor, blood from the portal vein [269].

Microvessel permeabilities to water and solutes are generally considered to be constants that do not change unless the vessels are injured by inflammatory sequelae or traumatized. In fact, considerable evidence has been collected demonstrating that vascular permeability is not necessarily a constant and can change in seconds in response to volume regulatory hormones, such as the atrial natriuretic peptides [46, 109].

1.6   TRANSENDOTHELIAL FILTRATION MODIFIES ARTERIOLAR, CAPILLARY AND VENULAR FUNCTION

An emerging body of evidence indicates that transendothelial filtration exerts important effects on vascular function (Figure 1.8). For example, the flow of fluid across interendothelial junctions, through the internal elastic lamina in the arteriolar wall, and between vascular smooth muscle cells suspended in a three-dimensional collagen matrix alters the shear forces acting on arterial smooth muscle cells [255, 256, 257, 273] and appears to play an important role in modulating vessel tone,

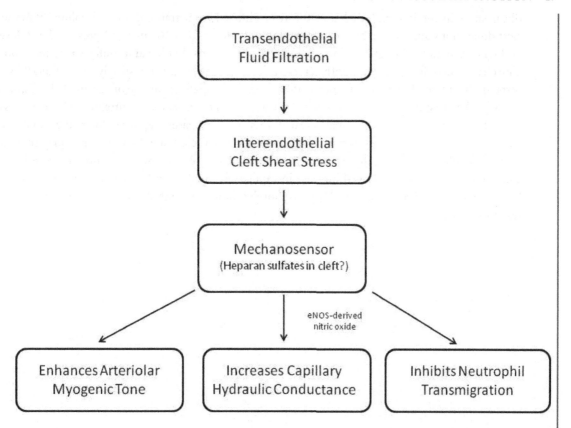

Figure 1.8: Alterations in transendothelial fluid flux modifies arteriolar tone, capillary hydraulic conductance, and neutrophil transmigration, effects that appear to involve shear sensing in the endothelial cleft. Heparan sulfates have been implicated as the mechanosensor for the interendothelial cleft flow changes. Shear stress-induced production of nitric oxide (NO) by endothelial nitric oxide synthase (eNOS) mediates the effects of enhanced transendothelial fluid flow to increase hydraulic conductance and inhibit neutrophil transmigration. Interestingly, shear stress on the apical surface (i.e., in the lumen) of endothelial cells also influences capillary hydraulic conductivity, suggesting that luminal and interendothelial cleft shear sensing may act cooperatively to promote increased conductance to transendothelial volume flow in response to vasodilation.

independent of shear effects on the apical glycocalyx on the endothelium [136]. Transendothelial flow has also been shown to inhibit neutrophil emigration across venular endothelial cells, an effect that appears to be related to endothelial cleft shear sensing and activation of endothelial nitric oxide synthase (eNOS) [30]. Increased capillary filtration also increases endothelial hydraulic conductivity, both in vitro and in vivo, an effect that appears to be related to eNOS activation [66, 135, 260].

Shear stress in the interendothelial clefts was estimated from transendothelial volume flows and cleft dimensions and ranged between 50 and 200 dyn/cm^2 for a 10 cm H$_2$O pressure head, levels far higher than noted in capillary lumens [30, 260]. It appears that heparan sulfates may function as a mechanosensor for the interendothelial cleft flow changes [66]. Interestingly, when fluid flowing through the interendothelial clefts percolates along the basolateral surface of endothelial cells, shear stress has been estimated at 10 dyn/cm^2 [255], a level that approximates intraluminal shear stress.

Since enhanced intraluminal shear forces also increases capillary hydraulic conductivity [134, 161, 196, 281, 282], luminal and interendothelial cleft shear stresses may play mutually supportive roles in reducing the resistance to transcapillary volume flow in response to vasodilation. The luminal shear stress-induced increase in capillary hydraulic conductivity involves the glycocalyx as a mechanosensor and occludin phosphorylation as a potential downstream effector of this response [161, 196].

CHAPTER 2

The Interstitium

Fluid flowing across the capillary walls must cross the interstitial spaces between parenchymal cells to gain access to the lymphatic vasculature for subsequent return to the vascular system (Figure 1.1). The interstitium does not simply represent a passive conduit system for the flux of fluid and solutes, but also functions as a highly dynamic and complex structure whose physical properties exert profound influences on fluid and solute exchange and the behavior of tissue cells. Capillary filtration drives fluid flow through the interstitium, which is essential for protein transport from the blood to parenchymal and interstitial cells, because these macromolecules are too large to readily diffuse through the ensemble of extracellular matrix components that fill the spaces between the vascular and lymphatic capillaries. In addition, this interstitial fluid flow also exerts important effects on tissue cells by shifting pericellular distribution of secreted proteins such as proteases, chemokines, and morphogens, thereby allowing for directed cell migration and guided cell/cell interactions during development and in pathologic states. The responses of tissue cells are also modified by mechanical forces exerted by flowing interstitial fluid, which exert shear forces on cell surfaces, pressure forces that deform cellular structures, or alters tethering forces at cell-matrix connections. Finally, interstitial flow may influence the formation of new lymphatic vessels in regenerating tissues. These topics will be reviewed in the next section.

2.1 COMPOSITION, STRUCTURE AND THREE-DIMENSIONAL ORGANIZATION OF THE EXTRACELLULAR MATRIX IN THE INTERSTITIAL SPACES

The composition and organization of the extracellular matrix determine the mechanical properties of the interstitium such as its strength, elasticity, and hydration [14, 39, 41, 42, 88, 166, 221]. The interstitium is composed mainly of collagen types I, III, and V, elastin, and glycosaminoglycans (mucopolysaccharides, such as hyaluronate and proteoglycans) which are mechanically entangled and cross-linked to form a gel-like reticulum reminiscent of a brush-pile in terms of its three-dimensional organization (Figure 1.1). These large polymeric molecules are synthesized by fibroblasts and released into the interstitial space. Fibroblasts also release a variety of enzymes that continuously degrade matrix components, such that complete turnover of the extracellular matrix occurs every 50 days [88]. The rate of synthesis is influenced by local conditions and hormonal factors (e.g., thyroid hormone).

Collagen represents a major structural component of interstitial spaces, functioning as a scaffold for support of the interstitial space and surrounding parenchymal cells (Figure 1.1). This extracellular matrix protein is formed into rod-like fibrils that are comprised of parallel linear arrays, which are then organized into bundles measuring several micrometers in diameter that exhibit tensile strengths that are approximately one-sixth of that measured for milled steel [39]. This high tensile strength is related to the formation of covalent intermolecular linkages produced by the oxidative deamination of specific lysine or hydroxylysine residues by the enzyme lysyl oxidase, a process that occurs extracellularly. Several structurally and genetically distinct collagens have been identified and collagen fibers found in the extracellular matrices of various tissues vary widely in terms of diameter, distribution, and relative contents of hydroxylysine, hydroxyproline, and glycosylated hydroxylysine residues.

Elastin is another major structural component of the extracellular matrix and is one of the most hydrophobic of all known proteins. Elastin-associated microfibrils are highly complex structures that appear as solid, branching and unbranching, fine and thick, rod-like fibers, can occur as concentric sheets, or can be arranged in three-dimensional meshworks. These microfibrils are found in tissues that undergo repetitive distension or passive lengthening movements and appear to confer elasticity to these structures.

While collagen fibrils and elastin provide much of the structural framework for tissues, the glycosaminoglycan constituents of the extracellular matrix play a major role in tissue hydration. These mucopolysaccharides are linear chains of disaccharide units that carry anionic charge sites. In the interstitial spaces, glycosaminoglycans exist as three-dimensional random coils that interact to produce a continuous network of intermeshing, entangled reticular structures that entrap water and resist compression by electrostatic repulsion of neighboring anionic sites and elastic recoil of the mechanically intertwined coils (Figure 1.1). Thus, the high fixed charge density of glycosaminoglycans establishes interstitial volume.

Proteoglycans form macromolecular assemblies that consist of a protein core that is long and rod-like, to which are attached numerous sulfated mucopolysaccharides by covalent bonds, yielding a bottle-brush configuration to the ensemble. The terminal ends of the proteoglycan cores attach by hydrogen bonds to form enormous aggregates of bottle-brush and random coil structures that become entangled in the randomly distributed collagen array in the tissue matrix to produce elastic, three-dimensional reticulum (Figure 1.1).

The physicochemical properties of the extracellular matrix are dynamic and apparently derive in large part from the behavior of the glycosaminoglycan molecules, of which hyaluronate is of principle importance. Thus, it is not surprising that the three-dimensional reticular structure of the extracellular matrix provides mechanical support for the tissues and provides a sponge-like continuum for containment of water and solutes. The gel-like properties of the interstitium limit the availability of free water for fluid flow, although rivulets of free fluid exist within the space. The flow of this free fluid in the interstitium, which is derived from capillary filtration, drives protein transport from the blood to parenchymal and interstitial cells (e.g., fibroblasts, dendritic cells, adipocytes, and inflammatory cells such as extravasated white blood cells and mast cells) because proteins are too large to readily diffuse the distances between capillaries. Dynamic stresses related to fluid flow in the interstitium also bestow a signaling function that serves as an important morphoregulator in tissue development, maintenance, and remodeling, as well as providing cues that allow interstitial cells to monitor the state of their surroundings, establish microenvironments, and guide immune cells towards draining lymphatic vessels (see below).

2.2   SOLUTE EXCLUSION AND OSMOTIC AMPLIFICATION IN THE EXTRACELLULAR MATRIX

The entangled nature of the fibrils allows the extracellular matrix to behave as if it were perforated by pores approximately 200–250 angstroms in diameter. Thus, one important property of the interstitium relates to the ability of the gel reticulum to exclude solutes from portions of the available gel water (Figure 2.1) [41, 42, 88, 280]. As a consequence, extravasated plasma proteins present in the interstitium are normally distributed in only a fraction of the interstitial volume because these large solute molecules cannot gain access to certain regions of the matrix meshwork (Figure 2.1). In other words, these large molecules are distributed into the matrix spaces that have dimensions larger than the solute (the accessible volume) and are excluded from microdomains with smaller dimension (the excluded volume). For a given large solute, the excluded volume varies inversely with the hydration state of the interstitium, which varies in accord with capillary filtration and the rate of lymphatic outflow. As fluid accumulates in the interstitium, the density of matrix fibers decreases, thereby increasing matrix porosity and reducing the fraction of tissue fluid from which the solute is excluded (Figure 2.1, left panel). On the other hand, tissue dehydration compacts the extracellular matrix and increases the excluded volume (Figure 2.1, right panel).

The functional significance of this exclusion phenomenon is related in part to its effect on oncotic pressure generated by plasma proteins distributed within the interstitial space. As fluid accumulates in the interstitium, tissue oncotic pressure falls, thereby reducing the balance of forces favoring capillary filtration and limiting edema formation. However, this reduction in tissue oncotic pressure is not simply due to dilution by the capillary filtrate, but also relates to the increased tissue water that becomes available for solute distribution on hydration that occurs secondary to the reduction in excluded volume on hydration. As a consequence, the exclusion phenomenon amplifies the oncotic buffering response to capillary filtration by further diluting the concentration of large solutes in the available gel water. The exclusion phenomenon also influences protein diffusion in the interstitial space because alterations in the distribution volume modify the effective surface area for, and the frictional resistance to, the diffusion of macromolecules (Figure 2.1). From the foregoing discussion, it should be apparent that exclusion properties for a given solute depend on its molecular size, the concentration of the various matrix elements, and hydration state of the interstitium. In addition, because both the permeating solutes and matrix elements are polyionic in nature, electrostatic interactions may influence the distribution spaces.

The fact that extracellular matrix molecules are mechanically entangled and cross-linked to yield a three-dimensional structure that behaves as if it were permeated by 200–250 angstrom diameter pores implies that migrating cells would encounter difficulty in passively traversing this

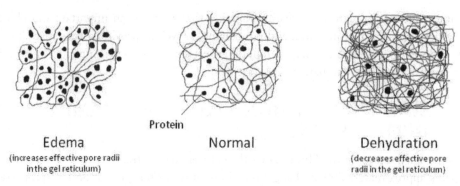

Edema
(increases effective pore radii
in the gel reticulum)

Protein

Normal

Dehydration
(decreases effective pore
radii in the gel reticulum)

Figure 2.1: Solute exclusion in the extracellular matrix gel reticulum varies with hydration state. With increased tissue hydration (left panel), as occurs in edema, the effective pore radii of the gel reticulum increase (i.e., matrix density decreases), resulting in a larger volume for distribution of solute. That is, a greater proportion of tissue water becomes available for protein distribution with matrix hydration. This phenomenon thereby amplifies the oncotic buffering response (reduction in tissue oncotic pressure) induced by filtration of protein-poor fluid, so that the decrease is tissue oncotic pressure is greater than would be expected from calculations based on dilution by the capillary filtrate, without a change in matrix density. Conversely, matrix dehydration (right panel) compacts the glycosaminoglycans in the tissue spaces, reducing the effective pore radii of the gel reticulum, resulting in a smaller proportion of tissue water that is available for protein distribution. The exclusion phenomenon also influences the rate of protein diffusion in the extracellular matrix since alterations in the volumes for protein distribution change the effective surface area for diffusion. Modified from reference [88].

space. Rather, it has been proposed that active mechanisms exist to promote changes in cell shape and amoeboid movement between the interstices of extracellular matrix molecules or involve protease-dependent degradation of extracellular matrix molecules to facilitate cell motility within this space. (This topic is briefly reviewed here because degradation of matrix components may contribute to enhanced capillary filtration and solute permeability in inflammation (see below)). For example, it has been recently shown that endothelial cells form vascular guidance tunnels during angiogenesis by a mechanism that is dependent on release of matrix type 1-metalloproteinase [50, 51, 52, 113, 248, 249]. In effect, these migrating endothelial cells digest a route of passage through the extracellular matrix, thereby creating physical spaces in the interstitium that serve as conduit pathways for both assembly and remodeling of tubular structures as well as recruitment of other cell types, such as pericytes, that are required for this morphogenetic process. Whether such processes are involved in the migration of immune cells, fibroblasts, and metastasizing tumor cells is unclear. However, penetration and transit through basement membranes appears to correlate with sites of focal matrix metalloproteinase activity, involves integrin-dependent adhesion steps, and occurs in regions where some matrix proteins are more sparsely expressed [72, 132, 140, 274]. Upon gaining access to the interstitial spaces, leukocytes migrate at much faster velocities than that noted for activated fibroblasts and tumor cells, implying mechanistic distinctions [189]. Activated fibroblasts and tumor cells employ pericellular proteolysis to migrate through interstitial extracellular matrix molecule barriers. Proteases bound to cell membranes co-cluster with specific integrins at contacts with substrates and cleave the extracellular matrix [72, 140]. In contrast, migrating leukocytes move through the extracellular matrix in an amoeboid manner independent of protease activity and integrins, tracking along collagen fibers and squeezing through narrow spaces within the matrix reticulum [22, 72, 132, 150, 189, 204]. More recent work has demonstrated differences in migratory behavior for monocytes versus neutrophils in the interstitial space, implying that individual leukocyte types may employ diverse processes to directionally navigate the extracellular space [132, 140].

2.3 COMPLIANCE AND HYDRAULIC CONDUCTANCE IN THE EXTRACELLULAR MATRIX

The compliance characteristics of the interstitium vary with hydration state and are of great importance in determining the hydrostatic forces operating across the capillary and lymphatic walls and thus fluid movements across these endothelial cell membranes [14, 87, 88, 141]. Under normal conditions, the volume conductance of the interstitium is very low because most of the interstitial fluid is immobilized in the tissue matrix. Tissue volume conductance is even lower under dehydrated conditions but increases dramatically with overhydration. From Figure 2.2, which depicts the compliance characteristics of the interstitial spaces, it is apparent that small changes in capillary filtration – and thus tissue volume – from normal levels produce very large increments in interstitial fluid pressure. However, there is an abrupt change in interstitial compliance when interstitial fluid volume increases by just 20% from its normal value of 25 ml/100g (i.e., to 30 ml/100g) (Figure 2.2). Although the mechanisms responsible for this abrupt shift to high compliance are uncertain, it is presumed to reflect a reversible disentanglement of hyaluronic acid chains and cross-link rupture [41, 42, 87, 88]. In addition, fibroblasts, which form attachments with the extracellular matrix, relax in response to mediators released during inflammation, and may contribute to the increase in compliance with increasing hydration under such conditions (Figure 2.3) [215, 279]. Since both hydraulic conductance and macromolecule exclusion of the interstitium are influenced by the degree of matrix hydration, an increase in interstitial fluid volume decreases excluded volume for proteins in this space and increases tissue hydraulic conductance, enhancing blood-to-lymph transport of fluid and macromolecules (Figures 2.1 and 2.2). Indeed, doubling interstitial fluid volume can increase its hydraulic conductance over a thousand fold.

The intestinal mucosal interstitium (and other organs with transporting epithelium) is rather distinctive in this regard in that its hydration state is not only influenced by capillary filtration, but also by net fluid absorption from the intestinal lumen. Indeed, when interstitial hydration occurs secondary to enhanced fluid absorption from the intestinal lumen, this may induce a larger change in interstitial compliance than produced by increased venous pressure (which increases capillary pressure and thus capillary filtration) because the congested microvasculature contributes to tissue

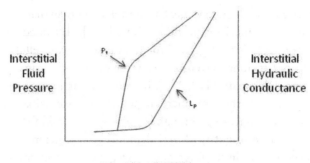

Figure 2.2: Relationship between interstitial fluid volume, interstitial fluid pressure, and interstitial hydraulic conductance. As interstitial fluid volume increases from its normal level, interstitial fluid pressure rises very quickly, owing to the low compliance characteristics of interstitium, and tissue hydraulic conductance is low. However, there is an abrupt change in interstitial compliance and conductance when interstitial volume increases by only 20% from its normal value. This abrupt shift to high compliance is thought to be due to a reversible disentanglement of hyaluronic acid chains and cross-link rupture. Modified from reference [87].

stiffness in the latter case [87]. These considerations may provide an explanation for the efficient transfer of chylomicrons (lipoprotein particles that are secreted by intestinal epithelial cells after a meal and range in size from 750 to 6000 A in diameter) through the interstitium, which should offer substantial resistance to their movement, while en route to the central lacteals of mucosal villi. Although it has been suggested that interstitial chylomicron movement may also be facilitated by inhomogeneities in the mucosal interstitial gel, which represent non-endothelialized channels that form on matrix expansion secondary to fluid absorption, ultrastructural studies have failed to demonstrate such pathways [87].

2.4   FLUID FLOW IN THE INTERSTITIUM MODIFIES THE FUNCTION OF TISSUE CELLS

Capillary filtration drives fluid flow into the interstitium, which not only assists transport of nutrients through tissues, but also plays important roles in tissue morphogenesis and remodeling, regulation of vascular function, inflammation and lymphedema, formation of new lymph vessels, tumor biology and immune cell trafficking (Figure 2.4) [89, 202, 224, 251, 252, 253]. The flow of fluid through the interstitium not only occurs through the three-dimensional ensemble of extracellular matrix molecules, but also around interstitial cells such as fibroblasts, extravasated immune cells, and adipocytes, as well as parenchymal cells comprising the tissue or organ (Figure 1.1). The flow of fluid derived from the capillary filtrate in the extravascular space occurs at a much slower velocity

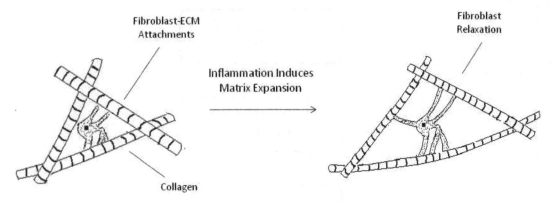

Figure 2.3: Fibroblasts form attachments to collagen fibrils in the extracellular matrix by an $\alpha2\beta1$-integrin-dependent mechanism, which confers cellular tension to the fibrous reticulum that restrains the hyaluronate/proteoglycan gel from taking up fluid and swelling. During inflammation, these cell attachments are loosened and/or fibroblasts relax, which allows the tissues to swell more readily and likely contributes to the abrupt transition in interstitial compliance that occurs when interstitial fluid volume increases by 20% or more. This effect to reduce matrix compaction attenuates the increase in interstitial fluid pressure that would normally occur, thereby reducing the effectiveness of this edema safety factor. In contrast, the presence of platelet-derived growth factor B-B, insulin, and prostaglandin F2α causes tissue cells to compact the gel reticulum, thereby increasing interstitial fluid pressure to more positive values. Modified from references [215] and [279].

(0.1–4 μm/s) than capillary luminal flow, owing to the high resistance offered by the extracellular matrix [37, 49]. However, this flow rate may increase 10-fold in edemagenic stress. It is important to note that interstitial flow occurs in all directions around the tissue cell-matrix interface, which may impart unique shear, pressure, and tethering force signals to tissue cells (Figure 1.1) [224]. Furthermore, interstitial fluid flow can also influence pericellular protein gradients, which may be particularly important for macromolecules that bind to matrix components, and are involved in directing cell migration (see below) [224].

The shear stress induced by capillary filtrate flowing through the extracellular matrix (0.005 to 0.015 dyn/cm^2) [224] is far smaller than that experienced in the capillary lumen, interendothelial clefts, or basal lamina, leaving it unclear as to whether such levels are sufficient to activate mechanosensors on tissue cells. Again, however, shear stress produced by interstitial flow may increase dramatically secondary to enhanced transcapillary filtration during edemagenic stress. Moreover, recent work indicates that the low shear stress levels normally present with interstitial flow can induce chemokine expression by dendritic cells embedded in an in vivo extracellular matrix construct [265]. Several other examples clearly indicate that interstitial flow exerts important effects on tissue function (Figure 2.4). In cartilage, interstitial flow has been shown to promote collagen

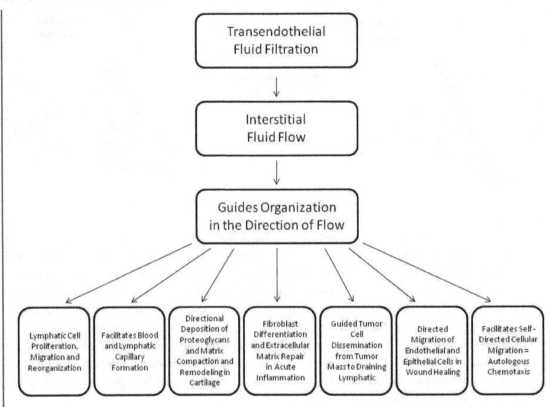

Figure 2.4: Capillary filtration drives flow into the interstitium, which not only assists transport of nutrients through tissues, but also plays an important role in guiding the organization of a wide variety of cellular processes in the direction of interstitial fluid flow. These include effects of flow in the interstitial space to influence tissue morphogenesis and remodeling, regulate vascular function, inflammation and lymphedema, direct the formation of new lymphatic vessels, guide tumor cell dissemination to lymph nodes, and facilitate immune cell trafficking in a directed manner.

and proteoglycan synthesis and increase chondrocyte metabolism [131, 210]. Interestingly, interstitial flow induces proteoglycan deposition and matrix fiber compaction in the direction of flow and guides remodeling by enhancing the transport of tissue inhibitor of metalloproteinases-1 [69, 77]. Interstitial fluid flow has also been suggested to direct the migration of endothelial and epithelial cells in wounded tissues [10, 160, 224] and appears to enhance blood and lymphatic capillary formation [26, 96, 97, 185, 235]. In regenerating skin, interstitial fluid channels form prior to the development of lymphatic capillaries, while lymphatic cell migration, expression of vascular endothelial cell growth factor-C, and lymphatic network organization occur in the direction of lymph flow [26]. These results suggest that lymphatic cells use the fluid channels for network development by a mechanism directed by interstitial flow.

Inflammation induces large increases in capillary filtration secondary to endothelial barrier dysfunction (increased permeability, reduced reflection coefficient) and elevated microvascular hydrostatic pressure induced by arteriolar vasodilatation. The associated increase in interstitial fluid flow, when coupled with hydrolytic cleavage of extracellular matrix components (which reduces the frictional resistance to solute movement), enhances the delivery of differentiation factors released from infiltrating inflammatory cells to fibroblasts, causing them to differentiate and remodel the extracellular matrix [224]. High interstitial flows characteristic of inflammatory states have also been shown to induce autocrine upregulation of transforming growth factor-b1 in fibroblasts seeded into three-dimensional collagen matrices, inducing their differentiation into myofibroblasts which then increase collagen production and alignment in a flow-directed manner [183, 184, 186]. These results suggest that enhanced capillary filtration associated with inflammation increases interstitial fluid flow, which in turn provides an early directional cue for rapid matrix repair. It has also been suggested that this may explain the development of tissue fibrosis induced by prolonged inflammation and why fibrosis can occur in the absence of inflammatory cells, as occurs with idiopathic pulmonary fibrosis [198, 233, 264].

Conditions characterized by abnormally low interstitial fluid flow may also induce dramatic changes in cell function. For example, lymphedema is a condition characterized by low interstitial flow that can result from either congenital lymphatic malformations (primary lymphedema) or from downstream lymphatic blockage that occurs with lymph node resection or compression due to tumor growth (secondary lymphedema). This results in excessive accumulation of fluid in the extravascular compartment in the absence of normal interstitial convection patterns, pathologic changes that cause chronic swelling of the interstitial compartment and precipitate inflammation and extensive remodeling of the extracellular matrix (fibrosis), adipocyte growth, and lipid accumulation [193, 225]. These changes are exacerbated by the absence of normal immune trafficking in the affected tissue [224]. These findings underscore the importance of normal interstitial fluid flow in the maintenance of healthy tissue.

Interstitial fluid flow has also been shown to affect pericellular diffusion gradients for signaling molecules such as growth factors, including members of the VEGF, FGF, Wnt and TGF families, and immunosuppressive chemokines such as CCL19 and CCL21 (Figure 2.5) [71, 211, 224, 265]. Many of these molecules, which are secreted by and can act upon the same cell, also bind strongly to elements of the extracellular matrix. Asymmetric distribution of the morphogen or chemokine by interstitial convective flow delivery, especially when maintained in pericellular proximity by matrix binding, creates both liquid and solid-phase gradients that provide the secreting cell with more nuanced and directed control of its microenvironment (Figure 2.5) [71, 211, 224, 265]. Matrix binding allows establishment of an autologous gradient of morphogen or chemokine in the direction of interstitial flow that the cell can follow for directed movement using cues that these cells secrete. Although the directional bias in the distribution of these signaling molecules that is introduced by interstitial

Cell-released
chemotaxin or
morphogen gradient
with no ECM binding
and no interstitial flow

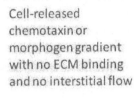

Cell-released chemotaxin or
morphogen gradient with no
ECM binding but with directional
interstitial flow → → →

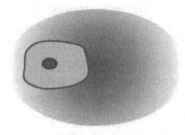

Cell-released chemotaxin or
morphogen gradient with binding
to and subsequent release from
ECM but with directional
interstitial flow → → →

Figure 2.5: In the presence of interstitial fluid flow, autologous pericellular gradients of self-secreted proteins (e.g., chemokines/morphogens) can develop in the direction of flow. The top panel depicts the chemotactic gradient development in the absence of interstitial flow, illustrating uniform pericellular distribution with chemokine or morphogen with concentrations decreasing radially and in a uniform manner as distance from the cell increases. However, this pattern of secreted protein distribution is altered by the presence of unidirectional flow of interstitial fluid, resulting in locally high concentrations downstream of the secreting cell that dissipate as distance from that cell increases (middle panel). This effect is magnified if the secreted chemokine or morphogen binds to extracellular matrix (ECM) components (bottom panel). To undergo chemotaxis in a directional manner, release of the chemotactic agent from the ECM, an event mediated by release of hydrolytic enzymes from the migrating cell, allows for washout of the chemotactic proteins that were secreted earlier by directional interstitial flow. This allows the locally high concentrations of the self-secreted chemokine or morphogen to be maintained by continued self-secretion and ECM binding/release at the leading edge of the migrating cell (bottom panel). This phenomenon of autocrine chemokine/morphogen gradient formation has been termed autologous chemotaxis and is thought to play an important role in leukocyte and tumor cell homing to draining lymphatics. Modified from references [212, 224, 253].

fluid flow is small owing to its low velocity, secretion of proteases from the same cell liberate the matrix-bound growth factor or morphogen, which amplifies the biasing effect of subtle interstitial flow and creates a stronger autocrine concentration gradient for directed motion (Figure 2.5) [71, 211, 224, 265]. This phenomenon, whereby cells receive directional cues while at the same time being the source of these prompting signals, has been termed autologous chemotaxis [71, 211, 224, 265]. Autologous chemotaxis has been demonstrated for VEGF directed capillary organization in vitro and more recently for leukocyte, dendritic cell, and tumor cell homing to draining lymphatics and lymph nodes [71, 96, 211, 224, 236, 239, 254, 265].

Interstitial fluid flow may also be exploited to enhance the targeted delivery of chemotherapeutic agents, especially given the promise of protein and antibody therapies, as well as synthetic drug carriers and delivery systems, including nanoparticles, liposomes and retroviruses, all of which encounter transport limitations owing to their large size [214, 254]. The steric hindrance effects of the mechanically entangled and cross-linked extracellular matrix may be overcome by the use of ultra-small nanoparticles (25 nm) to activate the complement cascade, thereby generating signals to stimulate dendritic cells to trigger adaptive immune responses [214]. In addition, coincident introduction of enzymes such as collagenase or hyaluronidase to disrupt the extracellular matrix of tumors would be expected to facilitate the interstitial delivery of chemotherapeutic agents by enhanced convective flow and reduced molecular sieving.

C H A P T E R   3

The Lymphatic Vasculature

3.1   ANATOMY AND NOMENCLATURE OF THE LYMPHATIC VASCULATURE

While initial study of the blood vascular system dates back to the sixth century BC, the lymphatic vasculature was not discovered until 1622 by Asellius. In stark contrast to the blood vasculature, the lymphatic circulatory system has been far less extensively studied. The reason for this is not its late discovery, but rather pertains to the prevailing misperception that the lymphatics represent a largely passive system for return of extravasated fluid and proteins to the systemic circulation and that no specific molecular markers had been identified to distinguish cells comprising this circulatory system from those in the blood vasculature until the last two decades [44]. As such, no common nomenclature for the lymphatic vasculature has developed, so one previously used for rat mesenteric lymphatics will be introduced next [231].

Interstitial fluid, formed from the extravasation of solute and fluid from the capillaries, enters blind-ended sacs composed only of an endothelial layer that is tethered to the interstitial matrix. These bulbous sacs (10–60 μm diameter) are called initial, or terminal, lymphatics. Initial lymphatics possess overlapping endothelial cells that behave collectively like a valve, only permitting unidirectional entry of fluid, solute, and cells into the lumen of these vessels (Figure 3.1). The fluid, thereafter referred to as lymph, next moves into lymphatic vessels of a similar diameter, termed microlymphatics (or lymphatic capillaries), consisting of an endothelial layer and basement membrane [267]. Since the phrase 'lymphatic capillary' is sometimes applied to initial lymphatics, and 'lymphatic capillaries' may be much larger than traditional capillaries carrying whole blood, this ambiguous phrase will be avoided hereafter. Microlymphatic vessels then carry lymph towards the larger collecting lymphatic vessels. Collecting lymphatics (50–200 μm diameter) are composed of endothelial cells, a basement membrane, lymphatic muscle cells, pericytes, and endothelial valves that prevent retrograde lymph flow [205, 271]. Pericytes do not envelope the smaller initial and microlymphatics [205]. Valves are present in collecting lymphatics and differ from the initial lymphatic 'valve' in that they consist of two modified endothelial cell leaflets that meet in the vessel lumen, not unlike bicuspid valves of the larger mammalian venous system. The lymphatic muscle layer is unique in that it possesses both tonic and phasic contractile activity [232, 271]. Phasic contractions of lymphatic muscle, referred to as spontaneous contractions, aid in propelling lymph along the intervalvular segments of the lymphatic vessel, called lymphangions. After passing through lymph nodes and then larger collecting lymphatic ducts, lymph is finally propelled to the thoracic duct, which empties into the left subclavian vein.

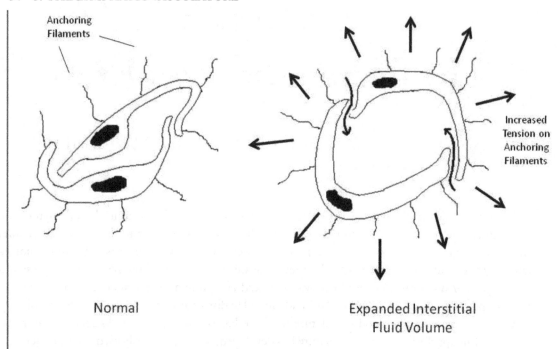

Normal

Expanded Interstitial
Fluid Volume

Figure 3.1: Expanding the interstitial fluid volume exerts radial tension on the anchoring filaments attached to endothelial cells of initial lymphatics, which increases luminal volume. This creates a small pressure difference that drives interstitial fluid into the lymphatic vessels. In addition, increased luminal diameter reduces the resistance to lymph flow. Note that the organization of anchoring filaments, coupled with the overlapping, interdigitating junctions between adjacent lymphatic endothelial cells creates a one-way valve system that prevents retrograde flow from the lymphatic capillary back into the interstitial space. Modified from [250].

The anatomy of collecting lymphatic vessels appears to be similar to that of comparable veins in that they both are low-pressure vessels vested with a muscle layer and intraluminal valves. In support of the theory of lymphatic development originally proposed by Sabin [227], one group has shown that lymphatic endothelial cells are derived directly from the cardinal vein [243]. Akin to the venules, numerous cytoplasmic vesicles have been reported in initial lymphatic endothelium [9, 33, 153, 154, 155, 188], but a role for these vesicles in solute uptake has not been elucidated fully at present. However, whether lymphatic vessels possess other similar features – such as a glycocalyx – has not been established.

Although the preceding description is accurate for the mesenteric lymphatic vasculature, it is important to recognize that lymphatic vessel morphology varies greatly between organs. Since it is beyond the scope of this section to thoroughly summarize these varying anatomical features, the reader is directed to several excellent detailed reviews for this information [14, 191, 232]. Two of these reviews [14, 232] summarize classical views on several aspects of the lymphatic vasculature, while the present chapter hopes to provide the reader with the most current perspectives, highlighting active areas of research and needed studies.

3.2   LYMPH FORMATION

Lymph formation refers to the entry of fluid and protein into the initial lymphatics. The mechanisms responsible for this process are poorly understood, but two main hypotheses have been proposed. The first suggests that an osmotic gradient becomes established across the initial lymphatic wall through sieving of protein that then generates its own convective flow by pulling in protein-containing interstitial fluid *against* a concentration gradient [34]. Very little, if any, experimental support exists for this unlikely theory despite its original appearance nearly four decades ago. Therefore, the following discussion will focus on the second hypothesis, which relies upon a hydrostatic pressure gradient to fill the initial lymphatics.

As stated before, the initial lymphatics possess overlapping endothelial cells tethered to the tissue (Figure 3.1). Thus, when the tissue becomes hydrated it swells and pulls apart the endothelial cells to form pores ~ 2 μm in diameter that act like a nonselective one-way valve, trapping fluid, solute, and cells passively (Figure 3.1) [266]. Considering their unique structure, one would arrive at the logical conclusion that a pressure gradient across the interstitium may drive fluid and solute accumulation within the initial lymphatics. Few studies on interstitial pressure gradients have been performed, but each proposes a gradient between 0.2–0.8 cmH$_2$O [99, 278]. At first glance this pressure gradient seems small, but others have calculated that a pressure head of only 0.12 cmH$_2$O is adequate to drive the capillary filtrate into the low resistance initial lymphatics [232]. The main problem with this hypothesis is that negative values of interstitial pressure are routinely measured [40, 91]. Significant overlap of the simultaneously measured interstitial and initial lymphatic pressures was observed [40, 100], depending on the superfusion solution and the time of measurement (immediately following exteriorization or 30 minutes later). Particularly interesting was that 30 minutes after exposure of the mesentery, superfused with oil to preserve natural tissue hydration, respective pressures in the interstitium and initial lymphatics were -0.2 and -0.25 mmHg [40]. Therefore, it is possible that a positive pressure gradient can allow fluid to enter the initial lymphatics even with a negative interstitial pressure. More current support for this hypothesis has been reported [182].

A passive interstitial pressure gradient, while sufficient, is not a complete description of every mechanism contributing to the formation of lymph. Pulsation of arteries was shown to aid in removal of interstitial tracer, which ceased after application of a steady arterial pressure [200]. Likewise, in the bat wing, cyclical dilation of the venules is a form of extrinsic pumping that also stimulates intrinsic spontaneous contractions of collecting lymphatics [63]. Other factors that increase local tissue pressure facilitate lymph formation such as respiration, muscle contraction (e.g., peristalsis, walking), elevated capillary filtration (e.g., venous hypertension, increased capillary permeability), and massage. Opposite to an increase in interstitial pressure, a variant of the hydrostatic pressure hypothesis posits that spontaneously contractile collecting lymphatic lymphangions, during their relaxation phase, are able to generate a suction force that draws interstitial fluid into the initial lymphatics [213]. Negative pressures produced by isolated bovine mesenteric collecting lymphatics under "low filling" states have been reported [78], but direct evidence for transmission of this suction to the initial lymphatics is needed.

Further support for the hydrostatic pressure gradient hypothesis is derived from studies demonstrating a positive correlation between interstitial pressure and lymph flow, which are discussed next.

3.3   INTERSTITIAL FLUID PRESSURE AND ITS INFLUENCE ON LYMPH FLOW

A convincing argument for an interstitial pressure gradient to drive lymph formation has been outlined in the previous section. The potential of the initial lymphatic lumen to collapse under a positive pressure difference is minimized by their unique anatomy. As noted above, initial lymphatic endothelial cells are tethered to the interstitium by anchoring filaments [156] responsible for holding the lumen open during conditions of increased tissue pressure or swelling, creating large ($\sim 2\ \mu$m diameter) interendothelial pores (Figure 3.1). The pores are a consequence of the punctate or "button"-like pattern of endothelial junctional adhesion proteins, in contrast to the contiguous expression of these molecules in blood vessel and collecting lymphatic endothelium [16]. Therefore, interstitial fluid is able to access the initial lymphatic lumen especially during edematous states when tissue pressure becomes positive.

Possibly as a result of the direct communication of the interstitial fluid and the lumen of the initial lymphatics, interstitial pressure and lymph flow are positively related. Several studies where tissue pressure was measured with the capsule technique provide direct evidence for this relationship [82, 262]. Figure 3.2A from Taylor and coworkers [263], shows that the rise of lymph flow in the dog hind leg is steepest at tissue pressures of $\sim$0–1 cmH$_2$O and attains a sustained maximal value when tissue pressure reaches 2 cmH$_2$O. The importance of this curve is that it maintains a constant interstitial volume due to the tight correlation between interstitial volume and interstitial pressure shown in Figure 3.2B [90]. Two mechanisms protecting against edema (i.e., edema safety factors) are evident from Figure 3.2, assuming that interstitial pressure is normally negative: 1) Since interstitial pressure must rise above 2 cmH$_2$O for lymph flow to plateau, large changes in interstitial pressure can be accommodated before edema develops, and 2) An elevated lymph flow will quickly return the interstitial volume back to normal levels, as long as the excess volume does not exceed the capacity of the lymphatic circulation. Thus, a small increase in interstitial volume greatly increases its pressure, promoting lymph flow that acts to restore the interstitial volume to normal.

3.4   LYMPHATIC SOLUTE PERMEABILITY

As summarized by Drinker [65], the main responsibility of the lymphatic circulaton is to be:

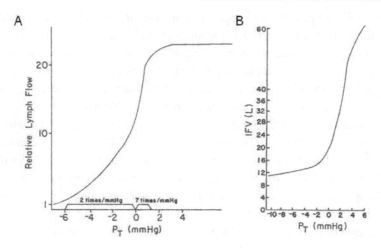

Figure 3.2: **A**, Relative increase in lymph flow versus tissue hydrostatic pressure (P_T, mmHg) during edema produced by intravascular infusion of Ringer's. **B**, Interstitial fluid volume (IFV in liters, L) versus P_T. Reprinted from reference [260], with permission.

"...engaged steadily in returning blood proteins to the blood, and that in the absence of normal lymph function these substances will accumulate extravascularly."

Therefore, because approximately 50% of the plasma proteins are filtered by the blood microvessels per day, and are *not reabsorbed by the venules* [159, 178], the lymphatic vasculature alone is left with the task of returning these proteins to the blood [84, 217]. Extravascular accumulation of plasma proteins, if unchecked, leads to the osmotic flow of water into the interstitium, producing edema. Further, Drinker [65] found that if thoracic duct lymph flow is diverted into a test tube, then the blood microvasculature "simply converts all the plasma to lymph." Such a circumstance is not compatible with life and illustrates the importance of a properly functioning lymphatic circulation that returns protein and fluid to the blood.

Many studies have probed the 'solute permeability' of initial lymphatics in a qualitative fashion, meaning that absolute values of permeability were not measured directly. Instead, after injection of colloid or radiolabeled protein, inferences were made from visualizing tracer uptake, analyzing downstream lymph samples, or viewing histological sections. As previously stated, both interendothelial pores [154] and transendothelial vesicles [9, 154] have been implicated in solute and fluid removal from the interstitium by initial lymphatics. The physiological role for vesicles in the lymphatic vasculature remains unknown, but their appearance may reflect the anatomic variation in lymphatic morphology [84]. Leak [154] favored the view that the interendothelial route was the major conduit for solute transport, whereas vesicles facilitated digestion of interstitial protein by the initial lymphatic endothelium. Generally, it is now believed that initial lymphatics are able to passively absorb particles, protein, cells, and fluid from the interstitium through large pores without regard for molecular size. Consequently, the lymph protein concentration of *initial* lymphatics probably approximates the protein concentration of the interstitium [261, 285]. Much controversy still surrounds whether the protein concentration of lymph from *collecting* lymphatics is equal to that of interstitial fluid; i.e., whether collecting lymphatics possess the ability to concentrate solute [29, 117, 258, 261].

Unlike initial lymphatic solute uptake, flux of solute across the collecting lymphatic walls depends on molecular size [167]. Studies performed by Mayerson [167] were initially aimed at answering questions regarding blood capillary permeability, using the lymphatic vaculature as a window into the interstitium, but became focused on how efficiently lymph is transported through the lymphatic circulation. By injecting a known amount of radiolabeled albumin directly into a canine limb collecting lymphatic duct (unknown diameter, but likely much larger than the peripheral collecting lymphatics, which average ~100 μm) and analyzing thoracic duct lymph, they were able to estimate the percent albumin lost across the vessel wall [201]. Not only is this method less sensitive than microfluorometric methods used today [108], but it almost certainly reflects the 'permeability' of the larger collecting lymphatic ducts and lymph nodes, not the prenodal microlymphatics or collecting lymphatic vessels. However, these studies were novel for the time and provided a first approximation of the size selectivity of the lymphatic ducts up to and including the thoracic duct. What Mayerson [167] discovered was that macromolecules equal to or greater than 6000 daltons (6 kDa) did not leave the larger ducts in 'great' quantity (albumin loss < 3%), while molecules smaller than 6 kDa escaped the vessels with ease. Simply put, large lymphatic ducts seemed to possess a relatively low permeability to large macromolecules and a higher permeability to molecules smaller than insulin.

Recently, a study of rat mesenteric collecting lymphatic solute flux largely confirmed Mayerson's observations in that the isolated vessel segments retained FITC-labeled dextran molecules of 4, 12, and 70 kDa progressively as molecular size increased [194]. Another group determined that the permeability to bovine serum albumin of cultured lymphatic endothelial cell tubes (~100 μm diameter) was $14\pm7 \times 10^{-7}$ cm·s^{-1} [208]. Conversely, a paper reporting rat mesenteric collecting lymphatic permeability to rat serum albumin (RSA) *in vivo* concluded that it did not significantly differ from venular RSA permeability ($3.5\pm1 \times 10^{-7}$ cm·s^{-1} vs. $4.0\pm1 \times 10^{-7}$ cm·s^{-1}, respectively), and that microlymphatic permeability to RSA was 12×10^{-7} cm·s^{-1} [231]. Three major implications are immediately apparent: 1) collecting lymphatics, at least in rat mesentery, exhibit a permeability to albumin no different from their embryological relative, the venules, supporting developmental work [243], 2) lymphatic endothelial cell tubes in culture that lack a basement membrane, lymphatic muscle, and pericytes presumably reflect microlymphatic versus collecting lymphatic permeability, warranting caution in data interpretation from cell culture studies, and 3) because lymph loses solute (and water) as it traverses the microlymphatics and collecting lymphatics, it is not identical to or representative of interstitial fluid. The last point must be emphasized, as lymph protein concentrations have been widely assumed to equal that of interstitial fluid. Indeed, lymph composition is further modified by nodal transit as evidenced by concentration differences in pre- vs postnodal lymph (which is the usual site for lymph collection for *in vivo* studies and often equated to interstitial fluid) [2, 3, 4, 5]. Scallan and Huxley [231] presented evidence supporting the hypothesis that lymph is concentrated by collecting lymphatics via loss of water over solute. When the total protein and albumin concentration of plasma, interstitium, and collecting lymphatic lymph were measured simultaneously, lymph protein concentrations were significantly greater than interstitial protein concentrations. Consequently, these data show that solute flux is directed from the vessel lumen to the interstitium; i.e., collecting lymphatics leak solute (Figure 3.3). However, there is still a need for direct measures of collecting lymphatic hydraulic conductivity (or 'water permeability coefficient') to confirm that the concentration of solute occurs as a result of losing more water than solute from the vessel. The fact that lymph flow is inversely related to lymph protein concentration lends further support to this hypothesis [29].

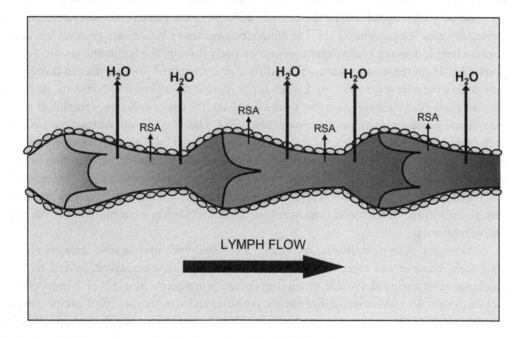

Figure 3.3:

But what does this mean in terms of edema? Lymphatic permeability may influence the effectiveness of the lymphatic edema safety factor given that vasoactive substances *increasing* collecting lymphatic permeability (to solute or water) may facilitate edema formation if most collecting lymphatics possess solute fluxes directed towards the interstitium. The implications of such findings in this newly revived area of research are that we will have to modify the conventional understanding of how lymphatic physiology affects fluid homeostasis.

3.5   PROPULSION OF LYMPH BY THE LYMPHATIC MUSCLE PUMP

The field of research examining the functional and molecular control of lymphatic contractility has been productive in the past two decades. One key discovery is that lymphatic muscle may act as a functional hybrid between smooth muscle and cardiac muscle because it contains molecular machinery from both cell types [181]. This has stimulated new hypotheses about how collecting lymphatics are able to independently regulate both tonic and phasic contractions.

Hydrostatic pressure increases progressively as lymph moves downstream into larger vessels of the lymphatic vasculature. On the contrary, peripheral veins experience a greater hydrostatic pressure than the downstream central veins, especially when standing, owing to the effects of gravity. Restated, the pressure gradient produced by the heart, in addition to the extrinsic venous pump, provides the driving force for venous return [14]. The lymphatic vasculature has no such pressure head (i.e., *vis a tergo*) so lymph does not – and cannot – *drain* passively through the lymphatic vessels, but requires propulsion. However, under edematous conditions, the interstitial pressure rises so that lymph may flow down a pressure gradient [209]. Lymph is transported throughout the lymphatic vasculature by intrinsic phasic contractions generated by the lymphatic muscle of collecting lymphatics that, along with valves, are necessary for unidirectional lymph flow. The spontaneous contractions are analogous to the cardiac contraction cycle consisting of a contraction and relaxation phase, stroke volume, and ejection fraction [20]. For comparison, the measured ejection fraction and contraction frequency of rat mesenteric collecting lymphatics were ~67% and 6 min^{-1}, respectively [20]. Understanding the functional and molecular regulation of lymphatic spontaneous contractions is essential for developing therapeutic treatments for edema (and lymphedema) centered on augmenting contraction amplitude and/or frequency.

Since the walls of collecting lymphatics are vested with muscle cells, they are able to regulate their diameter and tone, therefore modulating lymph flow resistance. Several factors, both mechanical and chemical, are able to regulate collecting lymphatic tone [271]. Mechanical stimuli include lymph flow, shear stress, hydrostatic pressure, and temperature. Hydrostatic pressure has been shown to elicit a myogenic response in collecting lymphatic muscle (measured during the relaxation phase) analogous to the arteriolar myogenic response [56]. In this study, an elevation in hydrostatic pressure induced constriction, thus reducing the end diastolic diameter of isolated collecting lymphatics, similar to the arteriolar myogenic response to pressure. Interestingly, addition of the neuropeptide substance P to the superfusion bath potentiated this effect of pressure on tone. Chemical factors influencing collecting lymphatic tone include neurotransmitters, neuropeptides, hormones, and metabolites [12, 58, 212]. For example, substance P increases basal collecting lymphatic tone [12, 58].

Similar to cardiac myocytes, length-tension curves have been determined for the perivascular muscle of collecting lymphatics, arterioles, and venules [20, 286, 287, 288]. Wall tension and stress derived from these curves were found to be lowest in rat mesenteric lymphatics, while mesenteric veins possessed higher tension and stress, and that of mesenteric arteries was the highest. Functionally, this makes sense in that arterioles, the resistance vessels, constrict to regulate pressure and flow; venules and lymphatics possess lower hydrostatic pressures reflecting their roles as capacitance vessels. The same group estimated the optimal preload (or hydrostatic pressure) for collecting lymphatic tone during peak active force to be ~5–13 cmH$_2$O, which compares well with measures of *in vivo* hydrostatic pressure [231, 290].

Other research has focused on the regulation of collecting lymphatic phasic activity. Functional studies demonstrated that lymph flow inhibits spontaneous contraction frequency and amplitude of both collecting lymphatic and thoracic duct isolated vessels [79]. However, the conclusion was that *in vivo* total lymph flow (defined as passive flow plus contraction-generated flow) would not be diminished as expected. Instead, lack of pumping activity was suggested as a mechanism to reduce the outflow resistance in the presence of high passive flows [79, 209].

Another vessel possessing spontaneous contractile activity besides the collecting lymphatics is the portal vein. Like the portal vein, collecting lymphatics were more sensitive to the rate of circumferential stretch than to the magnitude [57, 123]. These characteristics are consistent with developmental work showing that lymphatic endothelial cells are derived from the cardinal vein [243]. In experiments where isolated mesenteric collecting lymphatics were exposed to pressure ramps of different rates, bursts of increased contraction frequency were observed with increasing hydrostatic pressure ramps while inhibition of contraction frequency was observed on the ramps decreasing in pressure [57]. When the effects of substance P on collecting lymphatic sensitivity to stretch were assessed, both contraction amplitude and frequency were enhanced under basal conditions and in response to elevations in pressure [12, 58].

The molecular basis for lymphatic spontaneous contractions has just begun to be explored. Nitric oxide (NO) has been implicated in the modulation of collecting lymphatic spontaneous contractions [79]. Application of NO to the solution bathing isolated collecting lymphatics blunted spontaneous contraction frequency, amplitude, and ejection fraction in a fashion that imitated the effects of lymph flow. However, when the effect of elevated flow was investigated after the addition of the NO synthase inhibitor, L-NAME, NO did not fully explain the inhibition of contraction amplitude and frequency. NO involvement in the regulation of lymphatic spontaneous contractions is discussed in more detail in recent reviews [286, 287].

Importantly, calcium has been studied in the context of lymphatic phasic activity. Several different Ca^{2+} channels have been identified in lymphatic muscle, including L-type and T-type channels [101]. A hypothesis for spontaneous transient depolarizations (STDs) in the generation of spontaneous contractions has been proposed [270, 272]. At present, it is likely that summation of several STDs leads to a spontaneous contraction, with several ion channels appearing to be involved [270, 272]. However, it is at present unclear whether a single STD or summation of several STDs leads to a spontaneous contraction. Another hypothesis states that pacemaker cells generate a current which spreads throughout the lymphatic muscle [190], although the precise location of these cells was made difficult by the diffusion of current [138]. A newer study, however, has identified a subpopulation of lymphatic muscle cells that may act as pacemaker cells [168].

While evidence for the sympathetic innervation of lymphatic muscle is abundant, its role in altering spontaneous contractions has not been fully investigated. Several current reviews cover this topic more thoroughly [169, 268, 270, 286, 287].

The benefits of studying the mechanisms of lymphatic contractions are obvious – during edema, when the lymphatic vessels appear overwhelmed, an increase in pumping efficiency would be expected to recover proper fluid balance. While this work has contributed to our knowledge of regulation and generation of lymph flow, a complete molecular understanding necessitates future studies.

3.6   LYMPHANGIOGENESIS

Lymphatic vessel development *de novo* has been alluded to in previous sections where recent experiments support the hypothesis proposed by Sabin [227]; i.e., that lymphatic endothelial cells bud directly off of the cardinal vein to form the primitive lymphatic vasculature [243]. This is in contrast to the alternative hypothesis [103] that lymphatic endothelium is formed solely from lymphangioblasts residing in the tissue. Lymphangiogenesis, commonly defined as any event that stimulates lymphatic vessel growth [250], has been a highly topical area of investigation for nearly two decades. Some aspects of the molecular pathways have been identified, but an understanding of their importance to lymphangiogenesis is still being pursued.

One molecule important for the commitment of endothelial cells to the lymphatic phenotype is *Prox1*, which was first shown in mice to be necessary for the development of the lymphatic vasculature [277]. Homozygous deletion of *Prox1* resulted in embryos devoid of lymphatic vessels, which proved to be embryonically lethal [95]. Interestingly, an outbred line of $Prox1^{+/-}$ mice survived but developed adult-onset obesity, providing a link between malformed lymphatic vessels and visceral fat accumulation [95]. Very recently, it was demonstrated that down-regulation of *Prox1* leads to the dedifferentiation of lymphatic endothelium into blood vessel endothelium, its default phenotype [124]. Because of its importance in determining endothelial cell identity, *Prox1* is commonly used as a lymphatic marker, but is also expressed in the liver, pancreas, and brain. Another molecule used widely to identify lymphatic endothelium is lymphatic vessel endothelial hyaluronan receptor-1 (LYVE-1), which is expressed to a greater extent in initial versus collecting lymphatics in adult humans [267]. It is important to note that LYVE-1 is also expressed on infiltrating macrophages, as well as the sinusoidal endothelium of the liver and spleen, limiting its utility as a specific marker for lymphatic vessels [116]. Recent work suggests that this molecule does not play a role in lymphangiogenesis [116].

Much work has focused on vascular endothelial growth factors C and D (VEGF-C/D) as the prime regulators of lymphangiogenesis. Each molecule can bind to the receptors VEGFR-2, VEGFR-3, or to neuropilin-2 (Nrp2, expression limited to initial lymphatics) after proteolytic processing, which serves to increase receptor binding and specificity [127]. To further complicate the signaling mechanisms, VEGFR-2 and -3 can form homodimeric or heterodimeric receptor complexes to activate proliferation signals [61]. Additionally, *Prox1* upregulates VEGFR-3 expression [277], suggestive of a role for VEGF-C in development. Binding of VEGF-C to VEGFR-3 induces transduction of proliferation and survival signals in cultured cells [165], adult tissues [112, 119], and during development [129]. These cues then lead to migration and sprouting of lymphatic endothelial cells (LECs). Confirming its importance to lymphatic vasculogenesis, VEGF-C/VEGFR-3 signaling was necessary for the sprouting of LEC from the cardinal vein [129]. Surprisingly, mutation of the Nrp2 semaphorin receptor inhibits the formation of microlymphatics but not collecting lymphatics [284].

VEGF-C signaling is currently being explored for lymphedema therapy. Primary lymphedema, an inherited genetic disruption of the lymphatic vasculature, is known to result from mutations in the VEGFR-3 gene. The most common forms are Milroy's disease and Meige's disease, both of which are characterized by lymphatic vessel hypoplasia and fluid accumulation. A mouse model (*Chy* mouse) has been developed that mirrors Milroy's disease and is reversed by the over-expression of VEGF-C [130]. No animal models exist for Meige's disease, which has been estimated to account for nearly 94% of all primary lymphedema cases [220]. This has severely limited progress with regard to development of rational therapies. Secondary lymphedema is acquired after birth as a consequence of lymphatic vessel injury as a result of trauma, surgery, irradiation, infection, or cancer. This type of lymphedema usually results in the accumulation of protein-rich fluid in the tissues. However, breast cancer related lymphedema appears to be an exception to this general rule, being characterized by accumulation of a protein-poor exudate [17]. The mouse-tail model of secondary lymphedema has been examined with respect to VEGF-C therapy [38]. In this study, microlymphatic growth reversed the lymphedema and improved immune cell trafficking to normal levels. While these studies hold promise for therapeutic lymphangiogenesis, other molecules that may rescue lymphedema (discussed next) have yet to be fully investigated.

Like *Prox1*, the tyrosine kinase Syk as well as its adaptor protein SLP76 act to prevent the formation of anastomoses between lymphatics and the venous vessels. This notion is based on the fact that amino acid substitution mutations of these proteins promote the formation of communicating channels between the lymphatic and the venous systems [1]. Therefore, these two proteins are needed for the separation of the lymphatic and blood systems. The fact that Syk and SLP76 are also expressed in hematopoietic cells (but not endothelial cells) strongly suggests a novel role for hematopoietic cells in maintaining distinctly separate lymph and blood vasculatures.

Remodeling of the lymphatic network is a necessary component of vascular network maintenance. This includes pruning of microlymphatic networks, and recruitment of perivascular cells to collecting lymphatics (e.g., pericytes and muscle cells). The growth factor ephrinB2 appears to play a major role in these remodeling processes as evidenced by the fact that genetically engineered mice exhibiting defects in this protein possess hyperplasic lymphatics without intraluminal valves and are unable to prune microlymphatic networks [164].

The angiopoietin receptors, Tie1 and Tie2, are both expressed on lymphatic endothelial cells. Deletion of the gene for angiopoietin2 causes lymphatic hypoplasia, which was completely reversed by introducing the gene for angiopoietin1 in its place [74]. How angiopoietin (1 & 2) signaling ties into lymphangiogenesis is not well understood.

Thus far only the most pertinent molecules involved in lymphatic vascular development and growth have been discussed in detail. A large number of other molecules have been identified as potential mediators but their roles in lymphedema or lymphangiogenesis remain uncertain [11, 121]. Further elucidation of their function could prove useful in developing novel therapies for lymphatic diseases.

3.7 TUMORS AND LYMPHATIC METASTASIS

Traditionally, dissemination of cancer cells to organs distal to the primary tumor has been thought to be initiated only through tumor-associated blood vessels and not lymphatics within tumors. This concept likely arose as a consequence of the belief that intratumoral lymphatic vessels could not be penetrated by tumor cells (see [167]) or that they were collapsed in the high pressure environment within the tumor mass [157]. [A possible explanation for the former assumption is that blind reference to 'the lymphatics' in the literature – not specifying whether initial, micro-, or collecting – leads one to conclude erroneously that the properties of these three subtypes are identical. For example, initial lymphatics are almost freely permeable to solute and cells, while collecting lymphatics possess a relatively lower permeability.] However, it is now well established that initial and microlymphatic vessels invade tumors in a fashion similar to blood vessel neovascularization [244]. Two hypotheses have been proposed regarding tumor cell metastasis and the lymphatic vasculature. One favors passive absorption of free tumor cells into the lymphatic vasculature whereas the alternative hypothesis proposes that active signaling is involved [203]. Passive metastasis of tumor cells via the lymphatic circulation is certainly likely to occur, but current experiments are now probing the molecular mechanisms behind tumor cell metastasis, including tumor-associated lymphangiogenesis.

New research has provided support for the hypothesis invoking molecular regulation of tumor metastasis. An interesting mechanism has been described to explain chemotaxis of some cancer cells to the lymphatic circulation. Many breast cancer and melanoma cells express CCR7, the chemokine receptor for the ligand CCL21, constitutively secreted by the lymphatic vasculature. As a result, the cancer cells expressing CCR7 will chemotax towards the lymphatic circulation in response to CCL21 *in vitro* and *in vivo* [114]. In the same study, VEGF-C enhanced the secretion of CCL21 by cultured tumor cells. Coupled with the fact that many cancer cell types already express VEGF-C and -D [251], this is likely a primary mechanism by which cancer cells 'find' the lymphatic vasculature and promote lymph node metastasis. Further, a correlation between tumor-associated lymphangiogenesis and lymph node metastasis has been identified as a prognostic marker of disease [48]. However, whether or not lymphatic vessel density correlates with lymph node metastasis or a poor outcome has yet to be determined. Novel prognostic indicators of survival would be beneficial not only to the patient but also to the researcher as a guide to the mechanisms of tumor metastasis. Finally, other lymphatic growth factors have not enjoyed the attention VEGF-C has received [180], leading one to wonder whether other chemotactic or lymphangiogenic mechanisms exist.

3.8   CESSATION OF LYMPH FLOW AND ITS IMMUNOLOGICAL CONSEQUENCES

Continuous convection of interstitial fluid through the tissues and lymph through the lymphatic vessels is necessary for ensuring tissue health. When lymphatic vessels become dysfunctional due to either genetic or environmental processes, primary or secondary lymphedema will develop. A major consequence of any edema is tissue swelling, which serves to increase the distance molecules must diffuse to reach the cells in the tissue. Furthermore, immune cells monitoring the tissue have to cover longer distances before they reach the lymphatic circulation. In lymphedema where lymph flow is blocked, these immune cells become trapped in the tissue.

Antigen-presenting dendritic cells (APCs) recognize and bind antigens in the tissue to initiate the immune response, along with macrophages and natural killer cells. Once the antigen is bound, APCs express CCR7 allowing its chemotaxis towards the lymphatic vessels. Then the cell presents the antigen to T lymphocytes in the lymph nodes. The exact details and significance of these interactions are not fully understood at present [211, 287].

When APC migration to the lymphatic vasculature is hindered, as occurs during lymphedema or cancer, their presence in the tissue may exacerbate local tissue inflammation [13]. Once APCs bind antigen they begin to produce inflammatory chemokines to attract other leukocytes and immature APCs [211]. This process exacerbates the local tissue inflammation observed in patients with lymphedema [121].

Several aspects of immune function have yet to be explained fully in the context of lymphedema. For example, it is unclear how APCs gain access to the luminal side of lymphatic vessels, but probably occurs through the initial lymphatics, most likely by passive flux into the lumen via the flow of interstitial fluid through the one-way endothelial valves. It is also possible that intravasation may occur through the endothelial junctions of the larger collecting lymphatics in a process analogous to leukocyte transmigration across postcapillary venular endothelium. It is not clear whether the constitutive expression of CCL21 on the collecting lymphatic endothelium is required for chemotaxis of the APC to the lymph node [211].

CHAPTER 4

Pathophysiology of Edema Formation

Edema occurs when an excessive volume of fluid accumulates in the tissues, either within cells (cellular edema) or within the collagen-mucopolysaccharide matrix distributed in the interstitial spaces (interstitial edema) [14, 42, 62, 64, 87, 88, 141, 215, 247, 279]. Our focus is on swelling of the extracellular matrix or interstitial edema, which may occur as a result of aberrant changes in the pressures (hydrostatic and oncotic) acting across the microvascular walls, alterations in the molecular structures that comprise the barrier to fluid and solute flux in the endothelial wall that are manifest as changes in hydraulic conductivity and the osmotic reflection coefficient for plasma proteins, or alterations in the lymphatic outflow system, as predicted by examination of the Starling equation.

Excessive accumulation of interstitial fluid is generally viewed as detrimental to tissue function because edema formation increases the diffusion distance for oxygen and other nutrients, which may compromise cellular metabolism in the swollen tissue. For the same reason, edema formation also limits the diffusional removal of potentially toxic byproducts of cellular metabolism. These are especially important problems in the lungs, where pulmonary edema can significantly impair gas exchange. In some tissues, certain anatomical structures limit the expansion of the tissue spaces in response to edemagenic stress. For example, the kidneys are enveloped by a tough fibrous capsule, the brain is surrounded by the cranial vault, and skeletal muscles in the volar and anterior tibial compartments are encased in tight fascial sheaths. As a consequence of the inability of these tissues to readily expand their interstitial volume, relatively small increments in transcapillary fluid filtration induce large increases in interstitial fluid pressure. This, in turn, reduces the vascular transmural pressure gradient and physically compresses capillaries, thereby reducing nutritive tissue perfusion [120]. In the intestine, unrestrained transcapillary filtration leads to exudation of interstitial fluid into the gut lumen, a phenomenon referred to as filtration-secretion or secretory filtration [87]. Filtration-secretion may compromise the absorptive function of the delicate intestinal mucosa and appears to occur as a result of the formation of large channels between mucosal cells in the villous tips when interstitial fluid pressure increases by greater than 5 mmHg [87]. Ascites, or the pathologic accumulation of fluid in the peritoneal cavity, occurs in cirrhosis and is caused by fluid weeping from congested hepatic sinusoids secondary to elevated portal venous pressure [223]. Ascites can predispose afflicted individuals to peritoneal infections, hepatic hydrothorax, and abdominal wall hernias [223].

Hydrostatic edema refers to accumulation of excess interstitial fluid which results from elevated capillary hydrostatic pressure while permeability edema results from disruption of the physical structure of the pores in the microvascular membrane such that the barrier is less able to restrict the movement of macromolecules from the blood to interstitium. Lymphedema represents a third form and may result from impaired lymph pump activity, an increase in lymphatic permeability favoring protein flux from lumen to interstitial fluid, lymphatic obstruction (e.g., microfiliarisis), or surgical removal of lymph nodes, as occurs in the treatment of breast cancer. Destruction of extracellular matrix proteins, as occurs in inflammation secondary to the formation of reactive oxygen and nitrogen species and release of hydrolytic enzymes from infiltrating leukocytes, resident immune cells, and cells comprising the tissue parenchyma, alters the compliance characteristics of interstitial gel matrix such that interstitial fluid pressure fails to increase and oppose the movement of fluid. In addition, the tensional forces that are normally exerted by extracellular matrix proteins on the anchoring filaments (Figure 3.1) attached to lymphatic endothelial cells to facilitate lymphatic filling are diminished as a result of disrupted mechanical integrity [249]. Reductions in circulating plasma proteins, especially albumin, produce edema by decreasing plasma colloid osmotic pressure, and occurs in liver disease and severe malnutrition.

4.1   THE MARGIN OF SAFETY AGAINST EDEMA FORMATION – EDEMA SAFETY FACTORS

While increases in capillary pressure, reductions in plasma oncotic pressure, and/or disruption of endothelial barrier function are all accompanied by an increase in transmicrovascular filtration, the accumulation of fluid is resisted by a number of edema safety factors that work in concert to limit edema formation. This margin of safety against edema formation was first recognized in 1932 by Krogh and coworkers [148] as a means to explain why elevations in venous pressure by 10–15 mmHg failed to cause substantial accumulation of tissue fluid. Only when venous hypertension exceeded these levels did gross edema form, indicating that the margin of safety against edema formation could be overwhelmed. From the Starling equation (Equation (1.4)), one can readily see that increases in interstitial fluid pressure, reductions in tissue colloid osmotic pressure or microvascular surface area for exchange, or increases in lymph flow may all act to limit accumulation of excess fluid, and thus represent important edema safety factors against edema formation (Figures 4.1–4.5).

In addition to these basic compensatory mechanisms, the myogenic response to increased wall tension in arterioles and venous bulging constitute other edema safety factors in response to elevations in arterial or venous pressure in some tissues (Figure 4.1) [88]. Myogenic arteriolar vasoconstriction attenuates the rise in capillary pressure that might otherwise occur in response to arterial or venous hypertension, and also acts to reduce the microvascular surface area available for fluid exchange secondary to precapillary sphincter closure [55, 118, 131, 172]. When venous pressure is elevated, the volume of blood within postcapillary venules, larger venules and veins increases and bulge into the extravascular compartment, thereby raising tissue pressure. In effect, venous bulging stiffens the extracellular matrix by increasing tensional forces on the reticular fibers and fluid in

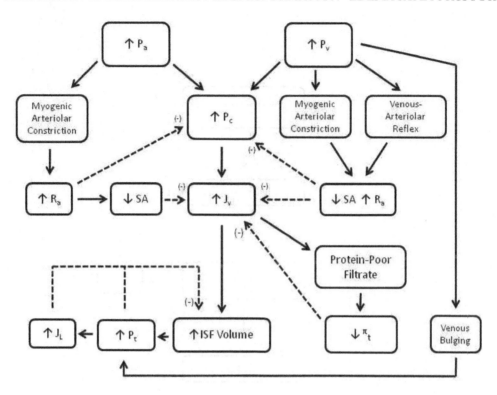

Figure 4.1: Mechanisms of enhanced transcapillary filtration in response to elevations in arterial or venous pressure. Elevations in arterial (P_a) or venous (P_v) pressure increase capillary pressure, which favors enhanced capillary filtration (J_v). The resulting increase in interstitial fluid volume raises tissue pressure (P_t) and thus lymph flow (J_L), both of which act as edema safety factors to oppose enhanced filtration. Because the capillary filtrate is protein-poor in composition, interstitial osmotic pressure decreases, an effect that is magnified by exclusion amplification. The latter changes also reduce the tendency for edema formation. Increased arterial or venous pressure also induces myogenic constriction of arterioles and precapillary sphincters, which raises arteriolar resistance (thereby minimizing the increase in capillary pressure) and reduces the microvascular surface area available for fluid exchange. Another factor that contributes to the margin of safety against edema formation when venous pressure is increased bulging of these highly compliant vessels, which stiffens the extracellular matrix, thereby elevating tissue pressure. Very large increases in venous pressure, as occurs in severe heart failure, may induce a "stretched pore phenomenon" to increase the effective pore radii in the microvascular membrane, thereby reducing its restrictive properties. This is manifested in the Starling equation as an increase in hydraulic conductance and a reduction in the osmotic reflection coefficient. The increase in permeability associated with the stretched pore phenomenon results in formation of a protein-rich filtrate that further facilitates transcapillary fluid movement.

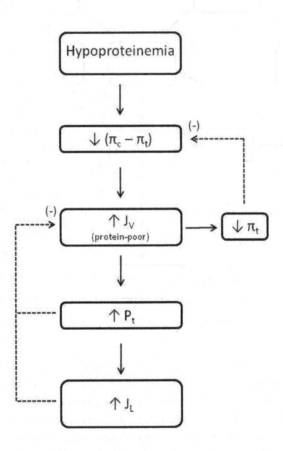

Figure 4.2: Hypoproteinemia reduces the effective colloid osmotic pressure gradient ($\pi_c - \pi_t$), resulting in increases in transcapillary fluid flux (J_V). The resulting increase in interstitial fluid volume raises interstitial fluid pressure (P_t) and thus lymph flow (J_L), changes that act to limit further accumulation of interstitial fluid (edema safety factors). Because the capillary filtrate is protein-poor in composition, interstitial fluid protein concentration is reduced, leading to a reduction in interstitial colloid osmotic pressure (π_t). This acts to oppose the effects of hypoproteinemia to reduce the transcapillary colloid osmotic pressure gradient and thus reduces transcapillary filtration and attenuates the tendency for edema formation.

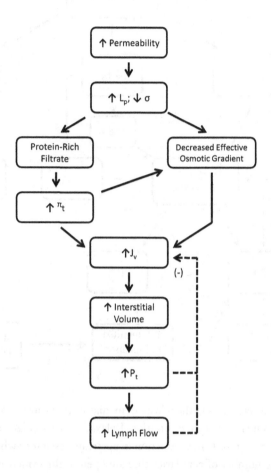

Figure 4.3: Increased microvascular permeability results in the formation of a protein-rich filtrate that raises interstitial colloid osmotic pressure (π_t), thereby reducing the effective colloid osmotic pressure gradient ($\sigma(\pi_c - \pi_t)$) acting across the microvascular wall. The increase in the diameter of large pores in the microvascular barrier that underlies the permeability increase also contributes to a reduction in the effective colloid osmotic pressure gradient by reducing the osmotic reflection coefficient for total plasma proteins (σ). As a consequence, of these changes, transcapillary filtration rate (J_V) is enhanced. The ensuing increase in interstitial fluid volume raises interstitial fluid pressure (P_t) and lymph flow (J_L), which act as edema safety factors to oppose further accumulation of interstitial fluid.

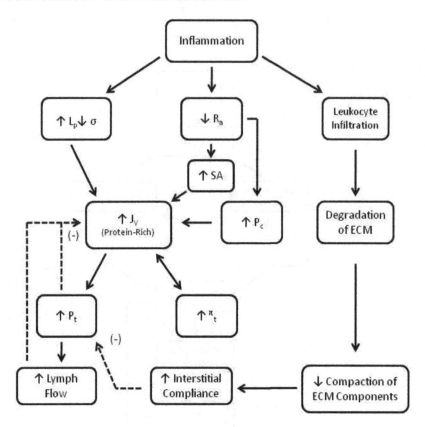

Figure 4.4: Inflammation results in the release of mediators that cause vasodilation, increase microvascular permeability, and induce leukocyte infiltration. Relaxation of vascular smooth muscle cells in arterioles and precapillary sphincters results in a reduction in upstream resistance which increases capillary pressure (P_c) and increases the number of capillaries that are open to flow (increasing surface area for exchange (SA)). These changes, coupled with increased effective pore radii in the microvascular barrier (which increases hydraulic conductivity, L_p, and reduces the osmotic reflection coefficient for total plasma proteins, σ), results in the formation of a protein-rich filtrate that increases interstitial fluid volume, interstitial fluid pressure (P_t), and tissue colloid osmotic pressure (π_t). Infiltrating leukocytes release a variety of reactive oxygen and nitrogen species, as well as hydrolytic enzymes, resulting in degradation of extracellular matrix proteins. In addition, mediator release induces fibroblast relaxation. Both degradation of extracellular matrix and fibroblast relaxation act to decrease the stiffness of the extracellular matrix (increased interstitial compliance), thereby attenuating the effect of increased interstitial fluid volume to increase interstitial fluid pressure, which is the driving force for lymph flow (J_L). Thus, the effectiveness of increased P_t and J_L as edema safety factors is compromised in inflammatory conditions characterized by leukocyte infiltration.

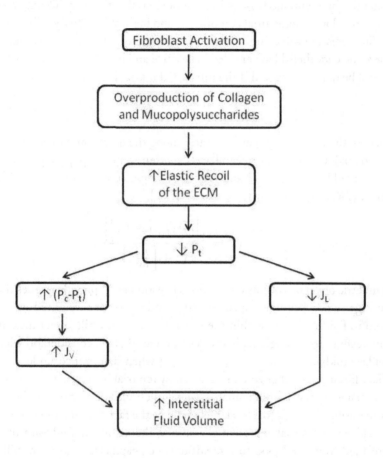

Figure 4.5: Myxedema is due to an accumulation of mucopolysaccharides secondary to overproduction of fibroblasts. This creates a suction force due to enhanced elastic recoil of the extracellular matrix that creates a high negative interstitial fluid pressure (P_t). This favors enhanced transcapillary filtration but may reduce lymphatic outflow, thereby producing edema. The dilution of interstitial protein concentration by enhanced filtration of protein-poor fluid reduces interstitial fluid colloid osmotic pressure, which acts as an edema safety factor to oppose the edemagenic effects of highly negative interstitial fluid pressure.

this space [88]. Finally, changes in excluded volume with increased transcapillary fluid filtration also comprise an important component of the margin of safety against swelling of the extracellular matrix compartment [88, 280].

From the aforementioned discussion, it is obvious that tissues exhibiting restrictive endothelial barrier properties, lowest interstitial compliance, and highest sensitivity of lymph flow to changes in interstitial fluid pressure will exhibit the greatest margin of safety against edema formation. Even in tissues where the endothelial barrier is less restrictive and lymphatic sensitivity is low, the margin of safety can still be quite substantial if the interstitial matrix is stiff.

4.2   VASOGENIC EDEMA

Disturbances in the vascular compartment are among the most common causes of interstitial edema (vasogenic edema) and result from capillary hypertension or hypoproteinemia. Capillary pressure (P_c) is determined by arterial (P_A) and venous (P_V) pressure and the ratio of pre- to postcapillary resistances (R_A/R_V) as shown by the equation [197]:

$$Pc = \frac{\left\{ \left[\frac{R_A}{R_V}(P_V) + P_A \right] \right\}}{\left\{ 1 + \left[\frac{R_A}{R_V} \right] \right\}} . \tag{4.1}$$

Using gravimetric or venous occlusion methods to estimate P_c provides values that range between 7 and 18 mmHg in a number of mammalian tissues and represent a weighted average for all microvessels involved in fluid exchange within the organ [142, 197], while direct measurements using micropuncture techniques in single capillaries yield values that are considerably higher (19–36 mmHg at the capillary midpoint) [28, 53, 54, 55, 85, 86] when determined under conditions where net transcapillary filtration is either zero or balanced by removal by lymph flow so that the tissue weight or volume remains constant (isogravimetric/isovolumetric). The discrepancy between values for capillary pressure using these approaches largely reflects the fact that gravimetric and venous occlusion methods yield estimates that represent pressure at the aggregate midpoint of vessels involved in filtration of fluid from the blood to interstitium (i.e., capillaries and postcapillary venules) under these conditions. Based on model analysis and the fact that direct micropuncture measurements of pressures within postcapillary venules range between 12 and 25 mmHg, it appears that the primary site of fluid filtration resides at or very near primary site of vascular compliance [142].

From Equation (4.1), it is apparent that capillary pressure rises when arterial or venous pressure increases and/or the pre- to postcapillary resistance ratio falls. Since arterial and venous pressure and the pre-to-postcapillary resistance ratio can be modified on a moment-to-moment basis in various physiologic (e.g., exercise) or pathologic conditions (e.g., inflammation) or following administration of vasoactive pharmaceutical agents, it might be expected that capillary pressure and thus transmicrovascular filtration rate can rapidly increase in accord with these changes. However, it has been suggested that capillary pressure may be tightly regulated in response to changes in arterial or venous pressure, by appropriate adjustments in pre- or postcapillary resistance, as a means to maintain a relatively constant interstitial fluid volume when any of these variables change [24, 55, 118, 173]. For example, because vascular smooth muscle in arterial and arteriolar walls contracts when exposed to elevated intravascular pressures, this myogenic response increases precapillary resistance and protects capillaries from a concomitant rise in their intravascular pressure. Conversely, when arterial pressure falls, myogenic tone is reduced in arterioles, decreasing their resistance to flow and maintaining capillary pressure. These observations suggest that capillary pressure may be regulated over the same range of pressure changes over which flow is autoregulated in a given organ. Indeed, from the relation:

$$P_c = P_V + Q R_V ,$$ (4.2)

one would predict that blood flow (Q) regulation would be perfectly coupled to the regulation of capillary pressure, assuming that venous pressure and resistance remain constant. However, an extensive analysis of changes in the pre-to-postcapillary resistance ratio and capillary pressure changes indicated that the effectiveness of flow and capillary pressure regulation are not always closely correlated, an effect that may be due to passive dimensional adjustments in capillaries and venules and rheological alterations in the blood flowing through these vessels as arterial pressure changes [55, 217]. In addition to the buffering effect of adjustments in the pre-to-postcapillary resistance ratio on capillary pressure, the influence of changes in capillary pressure induced by alterations in perfusion pressure are minimized by directionally opposite changes in the capillary filtration coefficient secondary to recruitment or derecruitment of perfused capillaries [172].

Similarly, changes in capillary pressure, and thus capillary filtration, are buffered when venous pressure is elevated [55, 125, 147]. At least two mechanisms account for this regulation of capillary pressure (Figure 4.1). Myogenic contraction of vascular smooth muscle in the walls of arterioles is elicited by transmission of the venous pressure increase to these upstream vessels [54, 55, 240]. A venous-arteriolar reflex has also been implicated in this response, wherein elevations in venous pressure activate antidromic impulses that are transmitted to nerve endings impinging on upstream arterioles, where neurotransmitter release elicits constriction [92, 234]. However, more recent work has challenged the importance of this mechanism versus the myogenic response [206]. It is important to note that capillary pressure, and thus capillary filtration, is not as well regulated in response to increases in venous pressure or resistance as when arterial pressure is altered [55, 144]. However, potential effects of increased venous pressure to reduce the capillary filtration coefficient may buffer the response to altered capillary pressure on transmicrovascular fluid movement, as outlined above.

While the aforementioned discussion focused on the effect of acute changes in venous pressure on the regulation of capillary pressure and transmicrovascular fluid movement and applies to most organs, the small intestinal vasculature may be unique in its response to chronic changes in venous pressure. Chronic intestinal venous hypertension induced by calibrated stenosis of the portal vein is associated with the development of a hyperdynamic circulation characterized by increased cardiac output, reduced intestinal vascular resistance, and increased intestinal blood flow [18, 19, 21, 147]. The latter changes result in a larger increase in intestinal capillary pressure than occurs during acute venous pressure elevations of the same magnitude and are associated with increases in the capillary filtration coefficient [147]. As a consequence, the increase in transcapillary filtration is much greater in chronic versus acute venous hypertension. The mechanisms responsible for the reduction in intestinal vascular resistance that account for the changes in capillary pressure and capillary filtration coefficient that lead to enhanced capillary filtration in chronic portal hypertension involve the formation of vasodilator substances and other factors and are reviewed elsewhere [18, 19, 21, 81, 115, 122, 146].

Capillary pressure is only modestly increased (~2 mmHg) in chronic arterial hypertension because the increase in arterial resistance that causes the rise in arterial blood pressure buffers transmission of the pressure increase to the capillary level [145]. Nevertheless, the associated increase in transmicrovascular filtration rate largely accounts for the elevated transcapillary escape rate of proteins noted in this disorder through convective coupling of fluid and protein flux. Elevated capillary pressure and filtration rate occur early in the course of development of diabetes mellitus and is thought to be an important stimulus for capillary basement membrane thickening, the ultrastructural hallmark of diabetic microangiopathy [27, 143]. Microvascular rarefaction, or loss of capillaries, has been reported to accompany the development of arterial hypertension, diabetes mellitus, and the metabolic syndrome [8, 27, 32, 70, 73, 143]. The attendant reductions in the surface area available for exchange may partially offset the effect of capillary hypertension to increase interstitial fluid volume in these conditions.

Very large increases in venous pressure may induce increments in capillary filtration far in excess of what would be predicted from the associated increase in capillary pressure. This is due to pressure-induced increases in microvascular permeability that are manifest in the Starling equation by increases in hydraulic conductivity and reductions in the osmotic reflection coefficient. For most organs, the permeability characteristics of the microvascular barrier to the exchange of fluid and lipid-insoluble solutes can be explained by the existence of large numbers of small pores with radii of 70 angstroms or less and a smaller number of large pores with radii in excess of 200 angstroms, with some models incorporating a third set of very small pores (< 10 angstroms in radius) to account for the diffusional flux of water. (Organs such as the liver, which have discontinuous capillaries characterized by large gaps between endothelial cells and reflection coefficients approaching 0.1, do not fit these models). Large increases in venous pressure are thought to enlarge these pores in microvascular wall, which is referred to as the stretched pore phenomenon [199, 218, 238]. Individual organs demonstrate a differential sensitivity to the effect of elevated venous pressure with regard to induction of the stretch pore phenomenon. For example, no increase in permeability occurs in microvessels of the feet during quiet standing, even though capillary pressure in the feet increases by more than 50 mmHg relative to values measured when supine, owing to the large hydrostatic column in arteries and veins. However, pulmonary capillaries may demonstrate a stretched pore phenomenon during conditions such as left ventricular failure, an effect that exacerbates pulmonary edema formation in this condition [199].

As noted above, myogenic constriction of arterioles in response to elevations in arterial or venous pressure constitutes an important safety factor against edema formation in hydrostatic edema by limiting the increase in capillary pressure and by reducing the number of perfused capillaries, and thus the available surface area for fluid filtration, that might otherwise occur in response to arterial or venous hypertension or increased venous resistance (Figure 4.1). However, it is important to note even modest increments in capillary pressure, which might appear to be small and inconsequential, can result in substantial increases in fluid filtration rates across the microvasculature. This is because normal net filtration pressure is quite small, averaging 0.15 mmHg for a prototypical body capillary. Thus, increasing capillary pressure by just 2 mmHg, as noted above in arterial hypertension, results in an initial 14-fold increase in fluid movement from the blood into the interstitium. Capillary hypertension results in the formation of a protein-poor ultrafiltrate that upon entry into the interstitial space raises interstitial fluid volume. Owing to the compliance characteristics of the interstitium, small increments in interstitial volume produce very large increases in tissue pressure, which effectively reduces the transcapillary hydrostatic pressure gradient, thereby limiting further accumulation of fluid (Figure 4.1). This effect is exacerbated in response to elevations in venous outflow pressure through the phenomenon of venous bulging. That is, the volume in veins increases immediately on elevation of venous pressure, which produces a coincident increase in interstitial pressure caused by expansion of engorged venules and veins into the interstitial spaces (Figure 4.1). In essence, venous engorgement shifts the interstitial compliance curve to the left, so that a smaller change in interstitial volume produces a larger increase in interstitial pressure. Increased interstitial fluid pressure increases lymph flow by three mechanisms. First, increased tissue pressure provides the driving pressure for flow into initial lymphatics. Second, increased pressure in the interstitial compartment creates radial tension on the anchoring filaments connecting the extracellular matrix to lymphatic endothelial cells, locally increasing initial lymphatic diameter and opening gaps between interdigitating and overlapping junctions between adjacent lymphatic endothelial cells (Figure 3.1). These tensional forces create a small, transient suction pressure for movement of interstitial fluid through enlarged gaps between adjacent endothelial cells, which act as a second, one-way valve system to ensure unidirectional flow from the interstitium into lymphatics. Third, as fluid moves into initial lymphatics, it increases volume in upstream lymphangions, promoting their contractile activity and lymph flow. The presence of valves between adjacent lymphangions assures one-way flow.

As noted above, capillary hypertension results in the movement of protein-poor fluid into the interstitial spaces, reducing the concentration of tissue proteins and decreasing tissue colloid osmotic pressure (Figure 4.1). This increases the effectiveness of the transcapillary oncotic pressure gradient ($\pi_c - \pi_t$) in opposing the hydrostatic gradient ($P_c - P_t$) favoring filtration. Because solute is excluded from a large portion of gel water in the extracellular matrix, the rapidity of the decrease in tissue protein concentration that occurs in response to increased interstitial fluid volume is enhanced, thereby augmenting the effectiveness of protein washdown as an edema safety factor. It is important to note that the effectiveness of decreases in tissue osmotic pressure as an edema safety factor is reduced in severe capillary hypertension, owing to the stretched-pore phenomenon discussed above, which increases convective-coupled protein transport into the tissue spaces.

4.3 HYPOPROTEINEMIA

Marked reductions in the circulating levels of proteins, especially albumin, is another cause of edema that relates to intravascular factors (Figure 4.2). Hypoproteinemia may result from rapid loss of proteins across a compromised glomerular barrier in diseased kidneys, impaired hepatic synthesis of plasma proteins in liver disease, severe malnutrition or protein-losing enteropathy (which limits the availability of substrate for protein synthesis), or from infusion of intravenous fluids lacking macromolecules. The ensuing reduction in the colloid osmotic pressure gradient ($\pi_c - \pi_t$), which favors reabsorption in the non-steady state and opposes the hydrostatic pressure gradient that favors filtration, induced by hypoproteinemia can result in a large transcapillary flux of protein-poor fluid into the interstitial spaces (Figure 4.2). Like capillary hypertension, this effect is opposed by elevations in tissue hydrostatic pressure, which increases lymph flow, both of which serve to limit the accumulation of tissue fluid (Figure 4.2). Enhanced capillary filtration also acts to dilute the concentration of proteins in the extracellular spaces, an effect that is magnified by increasing the accessible volume in the extracellular matrix gel (Figures 2.1 and 4.2). The ensuing reduction in interstitial colloid osmotic pressure acts to reduce net filtration pressure, thereby minimizing edema formation. Unlike the response to vascular hypertension, there is no stimulus for myogenic arteriolar vasoconstriction and venous bulging does not occur in hypoproteinemia, which reduces the margin of safety for edema formation in response to this edemagenic stress. As a consequence, tissues are less able to compensate for reductions in plasma colloid osmotic pressure that are equivalent to a given increase in capillary hydrostatic pressure.

4.4 PERMEABILITY EDEMA AND INFLAMMATION

Disruption of the microvascular barrier is a pathologic sequela in a large number of disease states, commonly accompanies trauma, and can be induced by a wide variety of endogenously produced mediators and pharmacologic agents. In the Starling equation (Equation (1.4)), this increase in permeability is manifest as a reduction in the osmotic reflection coefficient and/or an increase in hydraulic conductivity (Figure 4.3). Rapid reductions in the reflection coefficient decrease the effectiveness of the colloid osmotic pressure gradient in opposing filtration. The reduction in the restrictive properties of the endothelial barrier allows movement of a protein-rich filtrate into the tissue spaces, which increases interstitial colloid osmotic pressure (Figure 4.3). The resultant reduction in the colloid osmotic pressure gradient increases net filtration pressure, an effect that is exacerbated by the fact that many if not most of the mediators that increase microvascular permeability also act as vasodilators and reduce arteriolar resistance (Figure 4.4). As a consequence, capillary pressure is elevated, which further increases net filtration pressure. In addition, vasodilatation tends to recruit capillaries, thereby increasing microvascular surface area available for fluid and protein flux into the tissues. The latter change contributes to a further increase in the capillary filtration coefficient (which is equal to the hydraulic conductivity times surface area, $L_p S$), thereby magnifying the effect of increased net filtration pressure to promote volume flux. The marked enhancement in transcapillary fluid filtration results in increased convective transport of protein through the enlarged pores in the microvascular barrier (Figure 4.3). Under such conditions, the effect of increases in interstitial fluid pressure and lymph flow to provide a margin of safety against edema formation are rapidly overwhelmed and marked swelling of the interstitial spaces ensues.

Permeability edema is exacerbated in inflammatory states that are characterized by leukocyte infiltration into the tissues (Figure 4.4). Inflammation is a characteristic response to tissue injury and involves the release of a large number of mediators that not only increase microvessel permeability and cause vasodilatation, but also act to attract leukocytes to the damaged tissue (Figure 4.4). These phagocytic cells release a variety of hydrolytic enzymes as well as reactive oxygen and nitrogen species that degrade extracellular matrix components and the anchoring filaments that attach to lymphatic endothelial cells (Figures 3.1 and 4.4). This reduces the radial tension on the valve-like overlapping and interdigitating cell membranes at the interendothelial junctions in initial lymphatics, which may compromise lymphatic filling. Leukocyte-mediated disruption of the extracellular matrix components also increases interstitial compliance, which allows a larger volume of extracellular fluid to be accommodated within the matrix with little increase in interstitial fluid pressure, thereby attenuating the effectiveness of this edema safety factor. This effect is exacerbated by disruption of the connections fibroblasts form with collagen fibers in the interstitial spaces, which normally help compact the extracellular matrix by imposing tensional forces on these fibrillar components and restrain the gel matrix from taking up fluid and swelling (Figure 2.3). Extracellular matrix disruption thus produces a more compliant interstitium that limits the increase in interstitial fluid pressure for a given change in interstitial volume. Excluded volumes are also reduced by matrix degradation, an effect that increases effective interstitial colloid osmotic pressure. Extravasated proteins move more readily through the disrupted matrix, facilitating blood-to-lymph transport of these macromolecules.

4.5   NEUROGENIC INFLAMMATION

Neurogenic inflammation is characterized by leukosequestration, edema formation, and extravasation of plasma proteins following stimulation of sensory neurons. Sensory fibers release calcitonin gene-related peptide, substance P, and neurokinin A when stimulated. These proinflammatory molecules may act directly on the microvasculature to produce inflammation, but also appear to activate tissue mast cells, which augment the inflammatory response by release of their own complement of mediators.

4.6   MYXEDEMA

Myxedema is caused by suction of plasma filtrate into the tissue spaces that occurs as a result of overproduction of interstitial collagen and mucopolysaccharides by fibroblasts. Increasing the density of these extracellular matrix components augments the elastic recoil of the interstitial gel matrix and thereby producing a highly negative interstitial fluid pressure (Figure 4.5). As a consequence, lymph flow is reduced. Increased matrix density also increases the excluded volume, which acts to increase the effective interstitial colloid osmotic pressure. In effect, these changes create a suction force that accelerates fluid filtration and the development of edema. The most frequent manifestation of myxedema occurs in cases of hypothyroidism secondary to increased deposition of tissue matrix [149].

4.7   LYMPHEDEMA

Because lymphatic drainage represents the major route for removal of interstitial fluid (and macro-molecules) formed by capillary filtration, dysfunction of lymphatic vessels causes the development of edema and can exacerbate edema induced by other causes (Figure 4.6). Lymphedema occurs with physical obstruction of the lymphatic vessel lumen (either by extramural forces exerted by tumors or intraluminal obstruction by metastasizing tumor cells), destruction or regression of existing lymphatics, incompetence of the valves between lymphangions, paralysis of lymphatic muscle, reduced tissue motion, diminished arterial pulsations or venomotion, or by elevated venous pressure at the drainage points where lymphatics empty into the systemic blood circulation (Figure 4.6). Whatever the cause of lymphatic dysfunction, edema formation does not occur until lymph flow is reduced by 50%, all other factors being equal.

In the case of complete obstruction, lymph flow draining a tissue region falls to zero. Transcapillary filtration into this tissue region continues until interstitial pressure rises to equal net filtration pressure. As transcapillary volume flux decreases, the convective transport of protein from the vascular to interstitial compartments decreases. Since extravasated protein is not removed by the obstructed lymphatic, diffusional flux of protein continues until the concentration gradient is dissipated. At equilibrium, interstitial fluid pressure rises to microvascular hydrostatic pressure and interstitial colloid osmotic pressure equals plasma colloid osmotic pressure, yielding a net filtration pressure of zero. The affected tissue is characterized by large increases in water and protein content, fibrosis, and adipose cell deposition.

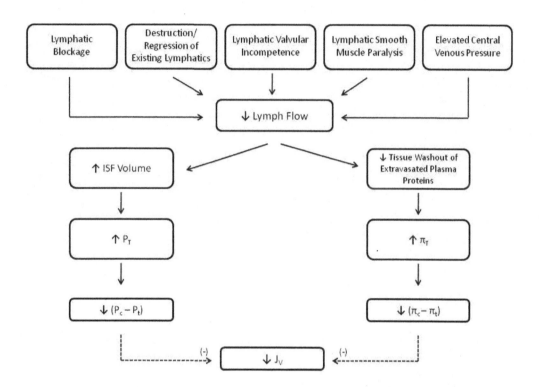

Figure 4.6: Lymphedema arises in response to a variety of conditions that result in reduced lymph flow. When lymphatic outflow (J_L) is completely occluded, interstitial fluid volume initially increases because capillary filtration (J_V) occurs until the interstitial Starling forces readjust to equal the Starling forces operating within the microvascular lumen. That is, because the occluded lymphatic represents the only pathway for net egress of extravasated plasma proteins from the tissue space, the decreased tissue washout of extravasated plasma proteins eventually results in dissipation of the diffusive gradient for protein flux from the blood to the tissue space and interstitial colloid osmotic pressure (π_t) rises until it equals plasma oncotic pressure (π_c). Likewise, continued capillary filtration in the absence of lymphatic outflow causes interstitial fluid pressure (P_t) to increase until it equals capillary pressure (P_c).

References

[1] Abtahian F, Guerriero A, Sebzda E, Lu MM, Zhou R, Mocsai A, Myers EE, Huang B, Jackson DG, Ferrari VA, Tybulewicz V, Lowell CA, Lepore JJ, Koretzky GA, and Kahn ML. Regulation of blood and lymphatic vascular separation by signaling proteins SLP-76 and Syk. *Science* 299: 247–251, 2003. DOI: 10.1126/science.1079477 43

[2] Adair TH, Moffatt DS, Paulsen AW, and Guyton AC. Quantitation of changes in lymph protein concentration during lymph node transit. *Am J Physiol* 243: H351: H359, 1982. 39

[3] Adair TH and Guyton AC. Modification of lymph by lymph nodes. II. Effect of increased lymph node venous blood pressure. *Am J Physiol* 245: H616-H622, 1983. 39

[4] Adair TH and Guyton AC. Modification of lymph by lymph nodes. III. Effect of increased lymph hydrostatic pressure. *Am J Physiol* 249: H777-H782, 1985. 39

[5] Adair TH, Montani JP, and Guyton AC. Modification of lymph by sheep caudal mediastinal node: effect of intranodal endotoxin. *J Appl Physiol* 57: 1597–1601, 1984. 39

[6] Adamson RH, Lenz JF, Zhang X, Adamson GN, Weinbaum S, and Curry FE. Oncotic pressures opposing filtration across non-fenestrated rat microvessels. *J Physiol* 557: 889–907, 2004. DOI: 10.1113/jphysiol.2003.058255 15

[7] Adamson RH, Huxley VH, Curry FE. Single capillary permeability to proteins having similar size but different charge. *Am J Physiol*, 254: H304-H312 , 1988. 11

[8] Agabiti-Rosei E. From macro-to microcirculation: benefits in hypertension and diabetes. *J Hypertens Suppl* 26: S15-S19, 2008. 56

[9] Albertine KH and O'Morchoe CC. Renal lymphatic ultrastructure and translymphatic transport. *Microvasc Res* 19: 338–351, 1980. DOI: 10.1016/0026-2862(80)90053-9 34, 37

[10] Albuquerque ML, Waters CM, Savla U, Schnaper HW, and Flozak AS. Shear stress enhances human endothelial cell wound closure in vitro. *Am J Physiol Heart Circ Physiol* 279: H293-H302, 2000. 28

[11] Alitalo K, Tammela T, and Petrova TV. Lymphangiogenesis in development and human disease. *Nature* 438: 946–953, 2005. DOI: 10.1038/nature04480 44

[12] Amerini S, Ziche M, Greiner ST, and Zawieja DC. Effects of substance P on mesenteric lymphatic contractility in the rat. *Lymphat Res Biol* 2: 2–10, 2004. DOI: 10.1089/1539685041690409 40, 41

[13] Angeli V, Ginhoux F, Llodra J, Quemeneur L, Frenette PS, Skobe M, Jessberger R, Merad M, and Randolph GJ. B cell-driven lymphangiogenesis in inflamed lymph nodes enhances dendritic cell mobilization. *Immunity* 24: 203–215, 2006. DOI: 10.1016/j.immuni.2006.01.003 45

[14] Aukland K and Reed RK. Interstitial-lymphatic mechanisms in the control of extracellular fluid volume. *Physiol Rev* 73: 1–78, 1993. 21, 25, 34, 35, 40, 47

[15] Baldwin AL and Winlove CP. Effects of perfusate composition on binding of ruthenium red and gold colloid to glycocalyx of rabbit aortic endothelium. *J Histochem Cytochem* 32: 259–266, 1984. 15

[16] Baluk P, Fuxe J, Hashizume H, Romano T, Lashnits E, Butz S, Vestweber D, Corada M, Molendini C, Dejana E, and McDonald DM. Functionally specialized junctions between endothelial cells of lymphatic vessels. *J Exp Med* 204: 2349–2362, 2007. DOI: 10.1084/jem.20062596 36

[17] Bates DO, Levick JR, and Mortimer PS. Change in macromolecular composition of interstitial fluid from swollen arms after breast cancer treatment, and its implications. *Clin Sci (Lond)* 85: 737–746, 1993. 43

[18] Benoit JN, Barrowman JA, Harper SL, Kvietys PR and Granger DN. Role of humoral factors in the intestinal hyperemia associated with chronic portal hypertension. *Am J Physiol* 247: G486-G493, 1984. 55, 56

[19] Benoit JN and Granger DN. Splanchnic hemodynamics in chronic portal hypertension. *Semin Liver Dis* 6: 287–298, 1986. DOI: 10.1055/s-2008-1040611 55, 56

[20] Benoit JN, Zawieja DC, Goodman AH, and Granger HJ. Characterization of intact mesenteric lymphatic pump and its responsiveness to acute edemagenic stress. *Am J Physiol* 257: H2059–2069, 1989. 40

[21] Benoit JN, Zimmerman B, Premen AJ, Go VL, and Granger DN. Role of glucagon in splanchnic hyperemia of chronic portal hypertension. *Am J Physiol* 251: G674-G677, 1986. 55, 56

[22] Bienvenu K, Harris N, and Granger DN. Modulation of leukocyte migration in mesenteric interstitium. *Am J Physiol* 36: H1573-H1577, 1994. 25

[23] Bingaman S, Huxley VH, and Rumbaut RE. Fluorescent dyes modify properties of proteins used in microvascular research. *Microcirc* 10: 221–232, 2003. DOI: 10.1080/mic.10.2.221.231 9

[24] Björnberg J, Gründe P-O, Maspers M, and Mellander S. Site of autoregulatory reactions in the vascular bed of cat skeletal muscle as determined with a new technique for segmental vascular resistance recordings. *Acta Physiol Scand* 133: 199–210, 1988. DOI: 10.1111/j.1748-1716.1988.tb08399.x 55

[25] Blum M, Müller UA, Höche A, Hunger-Dathe W, Stein G, and Strobel J. Flare measurement and albuminuria in type I diabetics. *Klin Monsatsbl Augenheilkd* 212: 80–83, 1998. DOI: 10.1055/s-2008-1034837 17

[26] Boardman KC, and Swartz MA. Interstitial Flow as a Guide for Lymphangiogenesis. *Circ Res* 92: 801–808, 2003. DOI: 10.1161/01.RES.0000065621.69843.49 28, 29

[27] Bohlen HG. Microvascular consequences of obesity and diabtes. In: *Handbook of Physiology: Microcirculation*, edited by Tuma RF, Duran WN, and Ley K, Chap 19, pp. 896–930, Elsevier, Amsterdam, 2008. 56

[28] Bohlen HG and Gore RW. Comparison of microvascular pressures and diameters in the innervated and denervated rat intestine. *Microvasc Res* 14: 251–264, 1977. DOI: 10.1016/0026-2862(77)90024-3 54

[29] Brace RA, Taylor AE, and Guyton AC. Time course of lymph protein concentration in the dog. *Microvasc Res* 14: 243–249, 1977. DOI: 10.1016/0026-2862(77)90023-1 38, 39

[30] Burns AR, Zhilan Z, Soubra SH, Chen J, Rumbaut RE. Transendothelial flow inhibits neutrophil transmigration through a nitric oxide-dependent mechanism: potential role for cleft shear stress. *Am J Physiol* 293: H2904-H2910, 2007. DOI: 10.1152/ajpheart.00871.2007 19, 20

[31] Calamita G. Aquaporins: highways for cells to recycle water with the outside world. *Biol Cell* 97: 351–353, 2005. DOI: 10.1042/BC20050017 17

[32] Carey RM. Pathophysiology of primary hypertension. In: *Handbook of Physiology: Microcirculation*, edited by Tuma RF, Duran WN, and Ley K, Chap 18, pp. 794–895, Elsevier, Amsterdam, 2008. 56

[33] Casley-Smith JR. The fine structure and functioning of tissue channels and lymphatics. *Lymphology* 13: 177–183, 1980. 34

[34] Casley-Smith JR. A theoretical support for the transport of macromolecules by osmotic flow across a leaky membrane against a concentration gradient. *Microvasc Res* 9: 43–48, 1975. DOI: 10.1016/0026-2862(75)90050-3 35

[35]   Cayrol R, Saikali P, and Vincent T. Effector functions of antiaquaporin-4 au-
toantibodies in neuromyelitis optica. *Ann NY Acad Sci*, 1173: 478–486, 2009.
DOI: 10.1111/j.1749-6632.2009.04871.x 17

[36] Celie JW, Beelen RH, and van den Born J. Heparan sulfate proteoglycans in extravasation:
assisting leukocyte guidance. *Front Biosci* 14: 4932–4949, 2009. 15

[37] Chary SR, and Jain RK. Direct measurement of interstitial convection and diffusion of albumin in normal and neoplastic tissues by fluorescence photobleaching. *Proc Natl Acad Sci USA*
86: 5385–5389, 1989. DOI: 10.1073/pnas.86.14.5385 27

[38] Cheung L, Han J, Beilhack A, Joshi S, Wilburn P, Dua A, An A, and Rockson SG. An experimental model for the study of lymphedema and its response to therapeutic lymphangiogenesis.
BioDrugs 20: 363–370, 2006. DOI: 10.2165/00063030-200620060-00007 43

[39] Chvapil M. *Physiology of Connective Tissue*, London, Butterworth, 1968. 21, 22

[40] Clough G and Smaje LH. Simultaneous measurement of pressure in the interstitium and the
terminal lymphatics of the cat mesentery. *J Physiol* 283: 457–468, 1978. 35

[41] Comper WD. Interstitium. *Edema* 9: 229, 1984. 21, 23, 25

[42] Comper WD and Laurent TC. Physiologic function of connective tissue polysaccharides.
Physiol. Rev 58: 255–315, 1978. 21, 23, 25, 47

[43]   Constantinescu AA, Vink H, and Spaan JA. Endothelial cell glycocalyx modulates immobilization of leukocytes at the endothelial surface. *Arterioscler Thromb Vasc Biol* 23: 1541–1547,
2003. DOI: 10.1161/01.ATV.0000085630.24353.3D 15

[44] Cueni LN and Detmar M. The lymphatic system in health and disease. *Lymphat Res Biol* 6:
109–122, 2008. DOI: 10.1089/lrb.2008.1008 33

[45]   Curry FE. Mechanics and thermodynamics of transcapillary exchange. In: *Microcirculation*,
Michel CC and Renkin EM ed. Baltimore: Williams & Wilkins, 309–374, 1984. 11, 15

[46] Curry FE. Atrial natriuretic peptide: an essential physiological regular of transvascular fluid, protein transport, and plasma volume. *J Clin Invest* 115: 1458–1461, 2005.
DOI: 10.1172/JCI25417 18

[47]   Curry FE, Huxley VH, and Adamson RH. Permeability of single capillaries to intermediate-
sized colored solutes. Am J Physiol, 245: H495-H505, 1983. 11

[48] Dadras SS, Lange-Asschenfeldt B, Velasco P, Nguyen L, Vora A, Muzikansky A, Jahnke
K, Hauschild A, Hirakawa S, Mihm MC, and Detmar M. Tumor lymphangiogenesis pre-
dicts melanoma metastasis to sentinel lymph nodes. *Mod Pathol* 18: 1232–1242, 2005.
DOI: 10.1038/modpathol.3800410 44

[49] Dafni H, Israely T, Bhujwalla ZM, Benjamin LE, and Neeman M. Overexpression of vascular endothelial growth factor 165 drives peritumor interstitial convection and induces lymphatic drain: magnetic resonance imaging, confocal microscopy, and histological tracking of triple-labeled albumin. *Cancer Res* 62: 6731–6739, 2002. 27

[50] Davis GE. Matricryptic sites control tissue injury responses in the cardiovascular system: Relationships to pattern recognition receptor regulated events. *J Mol Cel Cardiol* in press, 2009. DOI: 10.1016/j.yjmcc.2009.09.002 24

[51] Davis G and Senger DR. Endothelial Extracellular Matrix: Biosynthesis, Remodeling, and Functions During Vascular Morphogenesis and Neovessel Stabilization. *Circ Res* 97: 1093–1107, 2005. DOI: 10.1161/01.RES.0000191547.64391.e3 24

[52] Davis GE and Senger DR. Extracellular matrix mediates a molecular balance between vascular morphogenesis and regression. *Curr Opin Hematol* 15: 197–203, 2008. DOI: 10.1097/MOH.0b013e3282fcc321 24

[53] Davis MJ. Control of bat wing capillary pressure and blood flow during reduce perfusion pressure. *Am J Physiol* 255, H1114-H1129, 1988. 54

[54] Davis MJ. Microvascular control of capillary pressure during increases in local arterial and venous pressure. *Am J Physiol* 254: H772-H784, 1988. 54, 55

[55] Davis MJ, Hill MA, and Kuo L. Local regulation of microvascular perfusion. In: *Handbook of Physiology*, edited by Tuma RF, Duran WN, and Ley K, Chap 6, pp 161–284, Elsevier, Amsterdam, 2008. 48, 54, 55

[56] Davis MJ, Davis AM, Ku CW, and Gashev AA. Myogenic constriction and dilation of isolated lymphatic vessels. *Am J Physiol* 296: H293–302, 2009. DOI: 10.1152/ajpheart.01040.2008 40

[57] Davis MJ, Davis AM, Lane MM, Ku CW, and Gashev AA. Rate-sensitive contractile responses of lymphatic vessels to circumferential stretch. *J Physiol* 587: 165–182, 2009. DOI: 10.1113/jphysiol.2008.162438 41

[58] Davis MJ, Lane MM, Davis AM, Durtschi D, Zawieja DC, Muthuchamy M, and Gashev AA. Modulation of lymphatic muscle contractility by the neuropeptide substance P. *Am J Physiol* 295: H587–597, 2008. DOI: 10.1152/ajpheart.01029.2007 40, 41

[59]   Davis PF and Tripathi SC. Mechanical stress mechanisms and the cell: an endothelial paradigm. *Circ Res* 72: 239–245, 1993. 15

[60] Devuyst O, Ni J, and Verbavatz JM. Aquaporin-1 in the peritoneal membrane: implications for peritoneal dialysis and endothelial cell function. *Biol Cell* 97: 667–673, 2005. DOI: 10.1042/BC20040132 17

[61] Dixelius J, Makinen T, Wirzenius M, Karkkainen MJ, Wernstedt C, Alitalo K, and Claesson-Welsh L. Ligand-induced vascular endothelial growth factor receptor-3 (VEGFR-3) heterodimerization with VEGFR-2 in primary lymphatic endothelial cells regulates tyrosine phosphorylation sites. *J Biol Chem* 278: 40973–40979, 2003. DOI: 10.1074/jbc.M304499200 42

[62] Dongaonkar RM, Laine GA, Stewart RH, and Quick CM. Balance point characterization of interstitial fluid volume regulation. *Am J Physiol* 297: R6-R16, 2009. DOI: 10.1152/ajpregu.00097.2009 47

[63] Dongaonkar RM, Stewart RH, Laine GA, Davis MJ, Zawieja DC, and Quick CM. Venomotion modulates lymphatic pumping in the bat wing. *Am J Physiol* 296: H2015-H2021, 2009. DOI: 10.1152/ajpheart.00418.2008 35

[64] Dongaonkar RM, Quick CM, Stewart RH, Drake RE, Cox CS Jr, and Laine GA. Edemagenic gain and interstitial fluid volume regulation. *Am J Physiol* 294: R651-R659, 2008. DOI: 10.1152/ajpregu.00354.2007 47

[65] Drinker CK. The Functional Significance of the Lymphatic System: Harvey Lecture, December 16, 1937. *Bull N Y Acad Med* 14: 231–251, 1938. 36, 37

[66] Dull RO, Mecham I, and McJames S. Heparan sulfates mediate pressure-induced increase in lung endothelial hydraulic conductivity via nitric oxide-reactive oxygen species. *Am J Physiol* 292: L1452-L1458, 2007. DOI: 10.1152/ajplung.00376.2006 19, 20

[67] Dvorak AM and Feng D. The Vesiculo-Vacuolar Organelle (VVO): A New Endothelial Cell Permeability Organelle. *J Histochem Cytochem* 49: 419–32, 2001. 16

[68] Echtermeyer, F, Baciu PC, Caoncella S, and Goetinck PF. Syndecan-4 core protein is sufficient for the assembly of focal adhesions and actin stress fibers. *J Cell Sci* 112: 3433–3441, 1999. 15

[69] Evans RC and Quinn TM. Dynamic compression augments interstitial transport of a glucose-like solute in articular cartilage. *Biophys J* 91: 1541–1547, 2006. DOI: 10.1529/biophysj.105.080366 28

[70] Feihl F, Liaudet L, and Waeber B. The macrocirculation and microcirculation of hypertension. *Curr Hypertens Rep* 11: 182–189, 2009. DOI: 10.1007/s11906-009-0033-6 56

[71] Fleury ME, Boardman KC, and Swartz MA. Autologous morphogen gradients by subtle interstitial flow and matrix interactions. *Biophys J* 91: 113–121, 2006. DOI: 10.1529/biophysj.105.080192 29, 31

[72] Friedl P and Weigelin B. Interstitial leukocyte migration and immune function. *Nature Immunology* 9: 960–969, 2008. DOI: 10.1038/ni.f.212 24, 25

[73] Frisbee JC. Reduced nitric oxide bioavailability contributes to skeletal muscle microvessel rarefaction in the metabolic syndrome. *Am J Physiol* 289: R305-R306, 2005. DOI: 10.1152/ajpregu.00114.2005 56

[74] Gale NW, Thurston G, Hackett SF, Renard R, Wang Q, McClain J, Martin C, Witte C, Witte MH, Jackson D, Suri C, Campochiaro PA, Wiegand SJ, and Yancopoulos GD. Angiopoietin-2 is required for postnatal angiogenesis and lymphatic patterning, and only the latter role is rescued by Angiopoietin-1. *Dev Cell* 3: 411–423, 2002. DOI: 10.1016/S1534-5807(02)00217-4 44

[75] Gannon, BJ and Carati CJ. Endothelial distribution of the membrane water channel molecule aquaporin-1: implications for tissue and lymph fluid physiology? *Lymphat Res Biol.* 1: 55–66, 2003. DOI: 10.1089/15396850360495709 17

[76] Gao L and Lipowsky HH. Measurement of solute transport in the endothelial glycocalyx using indicator dilution techniques. *Ann Biomed Eng* 37: 1781–1795, 2009. DOI: 10.1007/s10439-009-9743-9 16

[77] Garcia AM, Lark MW, Trippel SB, and Grodzinsky AJ. Transport of tissue inhibitor of metalloproteinases-1 through cartilage: contributions of fluid flow and electrical migration. *J Orthop Res* 16: 734–742, 1998. DOI: 10.1002/jor.1100160616 28

[78] Gashev AA, Orlov RS, and Zawieja DC. [Contractions of the lymphangion under low filling conditions and the absence of stretching stimuli. The possibility of the sucking effect]. *Ross Fiziol Zh Im I M Sechenova* 87: 97–109, 2001. 36

[79] Gashev AA, Davis MJ, and Zawieja DC. Inhibition of the active lymph pump by flow in rat mesenteric lymphatics and thoracic duct. *J Physiol* 540: 1023–1037, 2002. DOI: 10.1113/jphysiol.2001.016642 41

[80] Gatinel D, Lebrun T, Le Toumelin P, and Chaine G. Aqueous flare induced by heparin-surface-modified poly(methyl methacrylate) and acrylic lenses implanted through the same-size incision in patients with diabetes. *J Cataract Refact Surg* 27: 855–860, 2001. DOI: 10.1016/S0886-3350(00)00814-2 17

[81] Gatta A, Bolognesi M, and Merkel C. Vasoactive factors and hemodynamic mechanisms in the pathophysiology of portal hypertension in cirrhosis. *Mol Aspects Med* 29: 119–129, 2008. DOI: 10.1016/j.mam.2007.09.006 56

[82] Gibson H and Gaar KA. Dynamics of the implanted capsule. *Fed Proc* 29: 319, 1970. 36

[83] Glinskii OV, Huxley VH, Turk JR, Deutscher SL, Quinn TP, Pienta KJ, and Glinsky VV. Continuous real time ex vivo epifluorescent video microscopy for the study of metastatic cancer cell interactions with microvascular endothelium. *Clin Exp Metastasis* 20: 451–458, 2003. DOI: 10.1023/A:1025449031136 11

[84] Gnepp DR. *Lymphatics*. New York: Raven Press, 1984. 37

[85] Gore RW. Pressures in cat mesenteric arterioles and capillaries during changes in systemic arterial blood pressure. *Circ Res* 34: 581–591, 1974. 54

[86] Gore RW and Bohlen HG. Pressure regulation in the microcirculation. *FASEB J* 34: 2031–2037, 1975. 54

[87] Granger ND, Kvietys PR, Perry MA, and Barrowman JA. The Microcirculation and Intestinal Transport. *Physiology of the Gastrointestinal Tract, Second Edition* 62: 1671–1697, 1987. 16, 25, 26, 47

[88] Granger HJ, Laine GA, Barnes GE, and Lewis RE. Dynamics and Control of Transmicrovascular Fluid Exchange. *Edema* 8: 189–224, 1984. 3, 21, 23, 24, 25, 47, 48, 54

[89] Griffith LG and Swartz MA. Capturing complex 3D tissue physiology in vitro. *Nat Rev Mol Cell Biol* 7: 211–224, 2006. DOI: 10.1038/nrm1858 26

[90] Guyton AC. Interstitial Fluid Pressure. II. Pressure-Volume Curves of Interstitial Space. *Circ Res* 16: 452–460, 1965. 36

[91] Guyton AC, Granger HJ, and Taylor AE. Interstitial fluid pressure. *Physiol Rev* 51: 527–563, 1971. 35

[92] Haddy FJ and Gilbert RP. The relation of a venous-arteriolar reflex to transmural pressure and resistance in small and large systemic vessels. *Circ Res* 4: 25–32, 1956. 55

[93] Haldenby, KA, Chappell DC, Winlove CP, Parker KH, and Firth JA. Focal and regional variations in the composition of the glycocalyx of large vessel endothelium. *J Vasc Res* 31: 2–9, 1994. 15

[94] Hamm S, Dehouch B, Kraus J, Wolburg-Buchholz K, Wolburg H, Risau W, Cecchelli R, Engelhardt B, and Dehouck MP. Astrocyte mediated modulation of blood-brain barrier permeability does not correlate with a loss of tight junction proteins from the cellular contacts. *Cell Tissue Res* 315: 157–166, 2004. DOI: 10.1007/s00441-003-0825-y 17

[95] Harvey NL, Srinivasan RS, Dillard ME, Johnson NC, Witte MH, Boyd K, Sleeman MW, and Oliver G. Lymphatic vascular defects promoted by Prox1 haploinsufficiency cause adult-onset obesity. *Nat Genet* 37: 1072–1081, 2005. DOI: 10.1038/ng1642 42

[96] Helm CL, Fleury ME, Zisch AH, Boschetti F, and Swartz MA. Synergy between interstitial flow and VEGF directs capillary morphogenesis in vitro through a gradient amplification mechanism. *Proc Natl Acad Sci USA* 102: 15779–15784, 2005. DOI: 10.1073/pnas.0503681102 28, 31

[97] Helm CL, Zisch A, and Swartz MA. Engineered blood and lymphatic capillaries in 3-D VEGF-fibrin-collagen matrices with interstitial flow. *Biotechnol Bioend* 96: 167–176, 2007. DOI: 10.1002/bit.21185 28

[98] Hillman NJ, Whittles CE, Pocock TM, Williams B, and Bates DO. Differential effects of vascular endothelial growth factor-C and placental growth factor-1 on the hydraulic conductivity of frog mesenteric capillaries. *J Vasc Res* 38: 176–186, 2001. DOI: 10.1159/000051044 9

[99] Hogan RD. *Lymph formation in the bat wing.* Kensington: Univ. of New South Wales, 1981. 35

[100] Hogan RD, Unthank JL. The initial lymphatics as sensors of interstitial fluid volume. *Microvasc Res* 31: 317–324, 1986. DOI: 10.1016/0026-2862(86)90020-8 35

[101] Hollywood MA, Cotton KD, Thornbury KD, and McHale NG. Tetrodotoxin-sensitive sodium current in sheep lymphatic smooth muscle. *J Physiol* 503: 13–20, 1997. DOI: 10.1111/j.1469-7793.1997.013bi.x 41

[102] Hu X and Weinbaum S. A new view of Starling's hypothesis at the ultra-structural level. *Microvasc Res* 58: 281–304, 1999. DOI: 10.1006/mvre.1999.2177 16

[103] Huntington GS and McClure CFW. The anatomy and development of the jugular lymph sac in the domestic cat (*Felis domestica*). *Am J Anat* 10: 177–312, 1910. 42

[104] Huxley VH and Curry FE. Albumin modulation of capillary permeability: Test of an absorption mechanism. *Am J Physiol*, 248: H264-H273, 1985. 9

[105] Huxley VH and Meyer Jr DJ. Capillary permeability: atrial peptide action is independent of "protein effect". *Am J Physiol* 259: H1351-H1356, 1990. 9

[106] Huxley VH and Williams DA. Basal and adenosine-mediated protein flux from isolated coronary arterioles. *Am J Physiol* 271: H1099-H1108, 1996. 11

[107] Huxley VH and William DA. Role of a glycocalyx on coronary arteriole permeability to proteins: evidence from enzyme treatments. *Am J Physiol* 278: H1177-H1185, 2000. 15

[108] Huxley VH, Curry FE, and Adamson RH. Quantitative fluorescence microscopy on single capillaries: a-lactalbumin transport. *Am J Physiol*, 252: H188-H197, 1987. 11, 38

[109] Huxley VH, Tucker VL, Verbug KM, and Freeman RH. Increased capillary hydraulic conductivity induced by atrial natriuretic peptide. *Circ. Res.* 60: 304–307, 1987. 18

[110] Huxley VH, McKay MK, Meyer Jr DJ, Williams DA, and Zhang R-S. Vasoactive hormones and autocrine activation of capillary exchange barrier function. *Blood Cells* 19: 309–324, 1993. 9

[111] Ichimura K, Stan RV, Kurihara H, and Sakai T. Glomerular endothelial cells form diaphragms during development and pathologic conditions. *J Am Soc Nephrol* 19: 1463–1471, 2008. DOI: 10.1681/ASN.2007101138 18

[112] Ikomi F, Kawai Y, Nakayama J, Ogiwara N, Sasaki K, Mizuno R, and Ohhashi T. Critical roles of VEGF-C-VEGF receptor 3 in reconnection of the collecting lymph vessels in mice. *Microcirculation* 15: 591–603, 2008. DOI: 10.1080/10739680701815538 43

[113] Iruela-Arispe ML and Davis GE. Cellular and Molecular Mechanisms of Vascular Lumen Formation. *Developmental Cell* 16: 222–231, 2009. DOI: 10.1016/j.devcel.2009.01.013 24

[114] Issa A, Le TX, Shoushtari AN, Shields JD, and Swartz MA. Vascular endothelial growth factor-C and C-C chemokine receptor 7 in tumor cell-lymphatic cross-talk promote invasive phenotype. *Cancer Res* 69: 349–357, 2009. DOI: 10.1158/0008-5472.CAN-08-1875 44

[115] Iwakiri Y and Groszmann RJ. Vascular endothelial dysfunctioin in cirrhosis. *J Hepatol* 46: 927–934, 2007. DOI: 10.1016/j.jhep.2007.02.006 56

[116] Jackson DG. Biology of the lymphatic marker LYVE-1 and applications in research into lymphatic trafficking and lymphangiogenesis. *APMIS* 112: 526–538, 2004. DOI: 10.1111/j.1600-0463.2004.apm11207-0811.x 42

[117] Jacobsson S and Kjellmer I. Flow and Protein Content of Lymph in Resting and Exercising Skeletal Muscle. *Acta Physiol Scand* 60: 278–285, 1964. DOI: 10.1111/j.1748-1716.1964.tb02889.x 38

[118] Järhult J and Mellander S. Autoregulation of capillary hydrostatic pressure in skeletal muscle during regional arterial hypo- and hypertension. *Acta Physiol Scand* 91: 32–41, 1974. DOI: 10.1111/j.1748-1716.1974.tb05654.x 48, 55

[119] Jeltsch M, Kaipainen A, Joukov V, Meng X, Lakso M, Rauvala H, Swartz M, Fukumura D, Jain RK, and Alitalo K. Hyperplasia of lymphatic vessels in VEGF-C transgenic mice. *Science* 276: 1423–1425, 1997. DOI: 10.1126/science.276.5317.1423 43

[120] Jerome SN, Akimitsu T, and Korthuis RJ. Leukocyte adhesion, edema, and development of postischemic capillary no-reflow. *Am J Physiol* 267: H1329–1336, 1994. 47

[121] Ji RC. Lymphatic endothelial cells, lymphedematous lymphangiogenesis, and molecular control of edema formation. *Lymphat Res Biol* 6: 123–137, 2008. DOI: 10.1089/lrb.2008.1005 44, 45

[122] Joh T, Granger DN, and Benoit JN. Endogenous vasoconstrictor tone in intestine of normal and portal hypertensive rats. *Am J Physiol* 264: H1135-H1143, 1993. 56

[123] Johansson B and Mellander S. Static and dynamic components in the vascular myogenic response to passive changes in length as revealed by electrical and mechanical recordings from the rat portal vein. *Circ Res* 36: 76–83, 1975. 41

[124] Johnson NC, Dillard ME, Baluk P, McDonald DM, Harvey NL, Frase SL, and Oliver G. Lymphatic endothelial cell identity is reversible and its maintenance requires Prox1 activity. *Genes Dev* 22: 3282–3291, 2008. DOI: 10.1101/gad.1727208 42

[125] Johnson PC. Effect of venous pressure on mean capillary pressure and vascular resistance in the intestine. *Circ Res* 16: 294–300, 1965. 55

[126] Jurgen W, VanTeeffelen GE, Constantinescu AA, Brands J, Spaan JAE, and Vink H. Bradykinin- and sodium nitroprusside-induced increases in capillary tube haematocrit in mouse cremaster muscle are associated with impaired glycocalyx barrier properties. *J Physiol*, 586: 3207–3218, 2008. DOI: 10.1113/jphysiol.2008.152975 15

[127] Jussila L and Alitalo K. Vascular growth factors and lymphangiogenesis. *Physiol Rev* 82: 673–700, 2002. 42

[128] Kamba T, Tam BY, Hashizume H, Haskell A, Sennino B, Mancuso MR, Norberg SM, O'Brien SM, Davis RB, Gowen LC, Anderson KD, Thurston G, Joho S, Springer ML, Kuo CJ, and McDonald DM. VEGF-dependent plasticity of fenestrated capillaries in the normal adult microvasculature. *Am J Physiol.* 290: H560–576, 2006. DOI: 10.1152/ajpheart.00133.2005 18

[129] Karkkainen MJ, Haiko P, Sainio K, Partanen J, Taipale J, Petrova TV, Jeltsch M, Jackson DG, Talikka M, Rauvala H, Betsholtz C, and Alitalo K. Vascular endothelial growth factor C is required for sprouting of the first lymphatic vessels from embryonic veins. *Nat Immunol* 5: 74–80, 2004. DOI: 10.1038/ni1013 43

[130] Karkkainen MJ, Saaristo A, Jussila L, Karila KA, Lawrence EC, Pajusola K, Bueler H, Eichmann A, Kauppinen R, Kettunen MI, Yla-Herttuala S, Finegold DN, Ferrell RE, and Alitalo K. A model for gene therapy of human hereditary lymphedema. *Proc Natl Acad Sci USA* 98: 12677–12682, 2001. DOI: 10.1073/pnas.221449198 43

[131] Kerin A, Patwari P, Kuettner K, Cole A, and Grodzinsky A. Molecular basis of osteoarthritis: biomechanical aspects. *Cell Mol Life Sci* 59: 27–35, 2002. DOI: 10.1007/s00018-002-8402-1 28, 48

[132] Khandoga AG, Khandoga A, Reichel CA, Bhiari P, Rehberg M, and Krombach F. *In Vivo* Imaging and Quantitative Analysis of Leukocyte Directional Migration and Polarization in Inflamed Tissue. *PLoS ONE* 4: e4693, 2009. DOI: 10.1371/journal.pone.0004693 24, 25

[133] Kikuchi K, Tancharoen S, Matsuda F, Biswas KK, Ito T, Morimoto Y, Oyama Y, Takenouchi K, Miura N, Arimura N, Nawa Y, Meng X, Shrestha B, Arimura S, Iwata M, Mera K, Sameshima H, Ohno Y, Maenosono R, Tajima Y, Uchikado H, Kuramoto T, Nakayama K, Shigemori M, Yoshida Y, Hashiguchi T, Maruyama I, and Kawahara KI. Edaravone attenuates cerebral ischemic injury by suppressing aquaporin-4. *Biochim Biophys Res Commun* Sept. 6 [Epub ahead of print], 2009. DOI: 10.1016/j.bbrc.2009.09.015 17

[134] Kim MH, Harris NR, and Tarbell JM. Regulation of capillary hydraulic conductivity in response to acute change in shear. *Am J Physiol* 289: H2126-H2136, 2005. DOI: 10.1152/ajpheart.01270.2004 20

[135] Kim MH, Harris NR, and Tarbell JM. Regulation of hydraulic conductivity in response to sustained changes in pressure. *Am J Physiol* 289: H2551-H2558, 2005. DOI: 10.1152/ajpheart.00602.2005 19

[136] Kim M, Harris NR, Korzich DH, and Tarbell JM. Control of the arteriolar myogenic response by transvascular fluid filtration. *Microvasc Res* 68: 30–37, 2004. DOI: 10.1016/j.mvr.2004.03.002 19

[137] Kimura M, Dietrich HH, Huxley VH, Reichner DR, and Dacey RG Jr. Measurement of hydraulic conductivity in isolated arterioles of rat brain cortex. *Am J Physiol* 264: H1788-H1797, 1993. 11

[138] Kirkpatrick CT and McHale NG. Electrical and mechanical activity of isolated lymphatic vessels [proceedings]. *J Physiol* 272: 33P-34P, 1977. 41

[139] Komarova YA, Mehta D, and Malik AB. Dual regulation of endothelial junctional permeability. *Sci STKE*, re8, 2007. DOI: 10.1126/stke.4122007re8 9

[140] Korpos E, WC, Song J, Hallmann R, and Sorokin L. Role of the extracellular matrix in lymphocyte migration. *Cell Tissue Res*, 330, 47–572010. 24, 25

[141] Korthuis RJ and Taylor AE. Interstitium and Lymphatic Techniques. *Microcirculatory Technology* 21: 317–342, 1986. 25, 47

[142] Korthuis RJ, Granger DN, and Taylor AE. A new method for estimating skeletal muscle capillary pressure. *Am J Physiol* 246: H880-H885, 1984. 54

[143] Korthuis RJ, Pitts VH, and Granger DN. Intestinal capillary filtration in experimental diabetes mellitus. *Am J Physiol* 253: G20-G25, 1987. 56

[144] Korthuis RJ, Granger DN, Townsley MI, and Taylor AE. Autoregulation of capillary pressure and filtration rate in isolate rat hindquarters. *Am J Physiol* 248: H835-H842, 1985. 55

[145] Korthuis RJ, Kerr CR, Townsley MI, and Taylor AE. Microvascular pressure, surface area, and permeability in isolated hindquarters of SHR. *Am J Physiol* 249: H498-H504, 1985. 56

[146] Korthuis RJ, Benoit JN, Kvietys PR, Townsley MI, Taylor AE, and Granger DN. Humoral factors may mediate increased rat hindquarter blood flow in portal hypertension. *Am J Physiol* 249: H827-H833, 1985. 56

[147] Korthuis RJ, Kinden DA, Brimer GE, Slattery KA, Stogsdill P, and Granger DN. Intestinal capillary filtration in acute and chronic portal hypertension. *Am J Physiol* 254: G339-G345, 1988. 55, 56

[148] Krogh A, Landis EM, Turner AH. The movement of fluid through the human capillary wall in relation to venous pressure and to the colloid osmotic pressure of the blood. *J Clin Invest* 11: 63–95, 1932. DOI: 10.1172/JCI100408 48

[149] Kwaku MP and Burman KD. Myxedema Coma. *J Intensive Care Med* 22: 224–231, 2007. DOI: 10.1177/0885066607301361 60

[150] Lämmermann T, Renkawitz J, Wu X, Hirsch K, Brakebusch C, Sixt M. Cdc42-dependent leading edge coordination is essential for interstitial dendritic cell migration. *Blood* 113: 5703–5710, 2009. DOI: 10.1182/blood-2008-11-191882 25

[151] Landis EM, Jonas L, Angevine M, and Erb W. The passage of fluid and protein through the human capillary wall during venous congestion. *J Clin Invest* 11: 717–734, 1932. DOI: 10.1172/JCI100445 2

[152] Landis EM and Gibbon JH. The effects of temperature and of tissue pressure on the movement of fluid through the human capillary wall. *J Clin Invest* 12: 105–138, 1933. DOI: 10.1172/JCI100482 2

[153] Leak LV. Electron microscopic observations on lymphatic capillaries and the structural components of the connective tissue-lymph interface. *Microvasc Res* 2: 361–391, 1970. DOI: 10.1016/0026-2862(70)90031-2 34

[154] Leak LV. Studies on the permeability of lymphatic capillaries. *J Cell Biol* 50: 300–323, 1971. DOI: 10.1083/jcb.50.2.300 34, 37

[155] Leak LV. The transport of exogenous peroxidase across the blood-tissue-lymph interface. *J Ultrastruct Res* 39: 24–42, 1972. DOI: 10.1016/S0022-5320(72)80004-2 34

[156] Leak LV and Burke JF. Ultrastructural studies on the lymphatic anchoring filaments. *J Cell Biol* 36: 129–149, 1968. DOI: 10.1083/jcb.36.1.129 36

[157] Leu AJ, Berk DA, Lymboussaki A, Alitalo K, and Jain RK. Absence of functional lymphatics within a murine sarcoma: a molecular and functional evaluation. *Cancer Res* 60: 4324–4327, 2000. 44

[158] Levick, JR. Revision of the Starling principle: new views of tissue fluid balance. *J Physiol* 557: 704, 2004. 16

[159] Levick JR. Capillary filtration-absorption balance reconsidered in light of dynamic extravascular factors. *Exp Physiol* 76: 825–857, 1991. 15, 37

[160] Li S, Huang NF, and Hsu S. Mechanotransduction in endothelial cell migration. *J Cell Biochem* 96: 1110–1126, 2005. DOI: 10.1002/jcb.20614 28

[161] Lopez-Quintero SV, Amaya R, Pahakis M, and Tarbell JM. The endothelial glycocalyx mediates shear-induced changes in hydraulic conductivity. *Am J Physiol* 296: H1451-H1456, 2009. DOI: 10.1152/ajpheart.00894.2008 20

[162] Luft JH. Fine structure of capillary and endocapillary layer as revealed by ruthenium red. *Microcirc Symp Fed Proc* 25: 1773–2783, 1966. 15

[163] Lum H, KA. and Roebuck KA. Oxidant stress and endothelial cell dysfunction. *Am J Physiol* 280: C719-C741, 2001. 9

[164] Makinen T, Adams RH, Bailey J, Lu Q, Ziemiecki A, Alitalo K, Klein R, and Wilkinson GA. PDZ interaction site in ephrinB2 is required for the remodeling of lymphatic vasculature. *Genes Dev* 19: 397–410, 2005. DOI: 10.1101/gad.330105 43

[165] Makinen T, Veikkola T, Mustjoki S, Karpanen T, Catimel B, Nice EC, Wise L, Mercer A, Kowalski H, Kerjaschki D, Stacker SA, Achen MG, and Alitalo K. Isolated lymphatic endothelial cells transduce growth, survival and migratory signals via the VEGF-C/D receptor VEGFR-3. *Embo J* 20: 4762–4773, 2001. DOI: 10.1093/emboj/20.17.4762 43

[166] Mathews, MB. *Connective Tissue, Macromolecular Structure and Evolutions*, Mol Biol Biochm Biophys 19: 1–318, 1975. 21

[167] Mayerson HS. *The physiologic importance of lymph*. Baltimore, MD: Williams & Wilkins, 1963. 38, 44

[168] McCloskey KD, Hollywood MA, Thornbury KD, Ward SM, and McHale NG. Kit-like immunopositive cells in sheep mesenteric lymphatic vessels. *Cell Tissue Res* 310: 77–84, 2002. DOI: 10.1007/s00441-002-0623-y 41

[169] McHale NG. *Nature of lymphatic innervation*. London: Portland Press, 1995. 42

[170] Mehlhorn U, Geissler HJ, Laine GA, and Allen SJ. Myocardial fluid exchange. *Eur J Cardiothorac Surg* 20: 1220–1230, 2001. DOI: 10.1016/S1010-7940(01)01031-4 15

[171] Mehta D and Malik AB. Signaling Mechanisms Regulating Endothelial Permeability. *Physiol Rev*, 86: 279–367, 2006. DOI: 10.1152/physrev.00012.2005 9

[172] Mellander S, and Johansson B. Control of resistance, exchange and capacitance functions in the peripheral circulation. *Pharmacol Rev* 20: 117–196, 1968. 48, 55

[173] Mellander S, Maspers M, Björnberg J, and Andersson LO. Autoregulation of capillary pressure and filtration in cat skeletal muscle in states of normal and reduced vascular tone. *Acta Physiol Scand* 129: 337–351, 1987. DOI: 10.1111/j.1748-1716.1987.tb08077.x 55

[174] Michel CC. Transport of macromolecules through microvascular wall. *Cardiovasc Res*, 32: 644–653, 1996. DOI: 10.1016/S0008-6363(96)00064-8 9, 11, 18

[175] Michel CC. Starling: the formulation of his hypothesis of microvascular fluid exchange and its significance after 100 years. *Exp Physiol* 82: 1–30, 1997. 16

[176] Michel CC and Curry FE. Microvascular Permeability *Physiol Rev* 79: 703 - 761; 2004. 16, 18

[177] Michel CC. Fluid exchange in the microcirculation, *J Physiol* 557: 701–702, 1999. DOI: 10.1113/jphysiol.2004.063511 16, 18

[178] Michel CC and Phillips ME. Steady-state fluid filtration at different capillary pressures in perfused frog mesenteric capillaries. *J Physiol* 388: 421–435, 1987. 37

[179] Mulivor AW and Lipowsky HH. Inflammation- and ischemia-induced shedding of venular glycocalyx. *Am J Physiol*. 286: H1672-H1680, 2004. DOI: 10.1152/ajpheart.00832.2003 15

[180] Mumprecht V and Detmar M. Lymphangiogenesis and cancer metastasis. *J Cell Mol Med* 13: 1405–1416, 2009. DOI: 10.1111/j.1582-4934.2009.00834.x 44

[181] Muthuchamy M, Gashev A, Boswell N, Dawson N, and Zawieja D. Molecular and functional analyses of the contractile apparatus in lymphatic muscle. *FASEB J* 17: 920–922, 2003. 40

[182] Negrini D and Fabbro MD. Subatmospheric pressure in the rabbit pleural lymphatic network. *J Physiol* 520: 761–769, 1999. DOI: 10.1111/j.1469-7793.1999.00761.x 35

[183] Ng CP and Swartz MA. Fibroblast alignment under interstitial fluid flow using a novel 3-D tissue culture model. *Am J Physiol* 288: H3016, 2005. DOI: 10.1152/ajpheart.01008.2002 29

[184] Ng CP and Swartz MA. Mechanisms of interstitial flow-induced remodeling of fibroblast-collagen cultures. *Ann Biomed Eng* 34: 446–454, 2006. DOI: 10.1007/s10439-005-9067-3 29

[185] Ng CP, Helm CL, and Swartz MA. Interstitial flow differentially stimulates blood and lymphatic endothelial cell morphogenesis in vitro. *Macrovasc Res* 68: 258–264, 2004. DOI: 10.1016/j.mvr.2004.08.002 28

[186] Ng CP, Hinz B, and Swartz MA. Interstitial fluid flow induces myofibroblast differentiation and collagen alignment in vitro. *J Cell Sci* 118: 4731–4739, 2005. DOI: 10.1242/jcs.02605 29

[187] Northover Am and Northover BJ. Involvement of protein kinase C in the control of microvascular permeability to colloidal carbon. *Inflammation Res* 39: 132–136, 1993. 9

[188] O'Morchoe CC, Jones WR, 3rd, Jarosz HM, O'Morchoe PJ, and Fox LM. Temperature dependence of protein transport across lymphatic endothelium in vitro. *J Cell Biol* 98: 629–640, 1984. DOI: 10.1083/jcb.98.2.629 34

[189] Ohaski KL, Tung DK, Wilson J, Zweifach BW, and Schmid-Schönbein GW. Transvascular and interstitial migration of neutrophils in rat mesentery. *Microcirculation* 3: 199–210, 1996. DOI: 10.3109/10739689609148289 25

[190] Ohhashi T, Azuma T, and Sakaguchi M. Active and passive mechanical characteristics of bovine mesenteric lymphatics. *Am J Physiol* 239: H88–95, 1980. 41

[191] Ohtani O and Ohtani Y. Organization and developmental aspects of lymphatic vessels. *Arch Histol Cytol* 71: 1–22, 2008. DOI: 10.1679/aohc.71.1 34

[192] Oki S, Desaki J, Taguchi Y, Matsuda Y, Shibata T, and Okumur H. Capillary changes with fenestrations in the contralateral soleus muscle of the rat following unilateral limb immobilization. *J Orthopaed Sci* 4: 28–31, 1999. DOI: 10.1007/s007760050070 18

[193] Olszewski WL. The lymphatic system in body homeostasis: physiological conditions. *Lymphat Res Biol* 1: 11–21, 2003. DOI: 10.1089/15396850360495655 29

[194] Ono N, Mizuno R, and Ohhashi T. Effective permeability of hydrophilic substances through walls of lymph vessels: roles of endothelial barrier. *Am J Physiol* 289: H1676–1682, 2005. DOI: 10.1152/ajpheart.01084.2004 38

[195] Pang Z and Tarbell JM. In vitro study of Starling's hypothesis in a cultured monolayer of bovine aortic endothelial cells. *J Vasc Res* 40: 351–358, 2003. DOI: 10.1159/000072699 15

[196] Pang Z, Antonetti DA, and Tarbell JM. Shear stress regulates HUVEC hydraulic conductivity by occludin phosphorylation. *Ann Biomed Eng* 33: 1536–1545, 2005. DOI: 10.1007/s10439-005-7786-0 20

[197] Pappenheimer JR, and Soto-Rivera A. Effective osmotic pressure of the plasma proteins and other quantities associated with the capillary circulation in the hindlimb of cats and dogs. *Am J Physiol* 152: 471–491, 1948. 54

[198] Pardo A and Selman M. Idiopathic pulmonary fibrosis: new insights in its pathogenesis. *Int J Biochem Cell Biol* 34: 1534–1538, 2002. DOI: 10.1016/S1357-2725(02)00091-2 29

[199] Parker RE, Roselli RJ, Harris TR, Brigham KL. Effects of graded increases in pulmonary vascular pressures on long fluid balance in unanesthetized sheep. *Circ Res* 49: 1164–1172, 1981. 56

[200] Parsons RJ and McMaster PD. The effect on the pulse upon the formation and flow of lymph. *J Exp Med* 68: 353–376, 1938. DOI: 10.1084/jem.68.3.353 35

[201] Patterson RM, Ballard CL, Wasserman K, and Mayerson HS. Lymphatic permeability to albumin. *Am J Physiol* 194: 120–124, 1958. 38

[202] Pedersen JA and Swartz MA. Mechanobiology in the third dimension. *Ann Biomed Eng* 33: 1469–1490, 2005. DOI: 10.1007/s10439-005-8159-4 26

[203] Pepper MS. Lymphangiogenesis and tumor metastasis: myth or reality? *Clin Cancer Res* 7: 462–468, 2001. 44

[204] Petri B, Phillipson M, and Kubes P. The physiology of leukocyte recruitment: an in vivo perspective. *J Immunol* 180: 6439–3446, 2008. 25

[205] Petrova TV, Karpanen T, Norrmen C, Mellor R, Tamakoshi T, Finegold D, Ferrell R, Kerjaschki D, Mortimer P, Yla-Herttuala S, Miura N, and Alitalo K. Defective valves and abnormal mural cell recruitment underlie lymphatic vascular failure in lymphedema distichiasis. *Nat Med* 10: 974–981, 2004. DOI: 10.1038/nm1094 33

[206] Ping P and Johnson PC. Role of myogenic response in enhancing autoregulation of flow during sympathetic nerve stimulation. *Am J Physiol* 263: H1177-H1184, 1992. 55

[207] Predescu D, Vogel SM, and Malik AB. Functional and morphological studies of protein transcytosis in continuous endothelia. *Am J Physiol* 287: L895-L901, 2004. DOI: 10.1152/ajplung.00075.2004 17

[208] Price GM, Chrobak KM, and Tien J. Effect of cyclic AMP on barrier function of human lymphatic microvascular tubes. *Microvasc Res* 76: 46–51, 2008. DOI: 10.1016/j.mvr.2008.02.003 38

[209] Quick CM, Venugopal AM, Gashev AA, Zawieja DC, and Stewart RH. Intrinsic pump-conduit behavior of lymphangions. *Am J Physiol* 292, R1510, R1518, 2007. DOI: 10.1152/ajpregu.00258.2006 40, 41

[210] Quinn TM, Grodzinsky AJ, Buschmann MD, Kim YJ, Hunziker EB. Mechanical compression alters proteoglycan deposition and matrix deformation around individual cells in cartilage explants. *J Cell Sci* 111: 573–583, 1998. 28

[211] Randolph GJ, Angeli V, and Swartz MA. Dendritic-cell trafficking to lymph nodes through lymphatic vessels. *Nat Rev Immunol* 5: 617–628, 2005. DOI: 10.1038/nri1670 29, 31, 45

[212] Reddy NP. Lymph circulation: physiology, pharmacology, and biomechanics. *Crit Rev Biomed Eng* 14: 45–91, 1986. 30, 40

[213] Reddy NP, Krouskop TA, and Newell PH, Jr. Biomechanics of a lymphatic vessel. *Blood Vessels* 12: 261–278, 1975. 36

[214] Reddy ST, van der Vlies AJ, Simeoni E, Angeli V, Randolph GJ, O'Neil CP, Lee LK, Swartz MA, and Hubbell JA. Exploiting lymphatic transport and complement activation in nanoparticle vaccines. *Nature Biotechnology* 25: 1159–1164, 2007. DOI: 10.1038/nbt1332 31

[215] Reed RK, Liden A, and Rubin K. Edema and fluid dynamics in connective tissue remodeling. *J Molec Cell Cardiol* in press, 2010. DOI: 10.1016/j.yjmcc.2009.06.023 25, 27, 47

[216] Reitsma S, Slaaf DW, Vink H, van Zandvoort MA, and oude Egbrink MG. The endothelial glycocalyx: composition, functions, and visualization. *Pflugers Arch* 454: 345–359, 2007. DOI: 10.1007/s00424-007-0212-8 15

[217] Renkin EM. Some consequences of capillary permeability to macromolecules: Starling's hypothesis reconsidered. *Am J Physiol* 250: H706-H710, 1986. 37, 55

[218] Rippe B, Haraldsson B, Folkow B. Evaluation of the 'stretched pore phenomenon' in isolated rat hindquarters. *Acta Physiol Scand* 125: 453–459, 1985. DOI: 10.1111/j.1748-1716.1985.tb07742.x 56

[219] Roberts, WG and Palade, GE. Neovasculature induced by vascular endothelial growth factor is fenestrated. *Cancer Res.* 57, 765–772, 1997. 18

[220] Rockson SG. Lymphedema. *Am J Med* 110: 288–295, 2001. DOI: 10.1016/S0002-9343(00)00727-0 43

[221] Roden L. Structure and metabolism of connective tissue proteoglycans. *The Biochemistry of Glycoproteins and Proteoglycans*, 267–371, 1980. 21

[222] Rumbaut RE, Harris N, Sial AJ, Huxley VH, and Granger DN. Leakage responses to L-NAME differ with fluorescent dye used to label albumin. *Am J Physiol.* 276: H333-H339, 1999. 9

[223] Runyon BA. Ascites and Spontaneous Bacterial Peritonitis. *Sleisenger & Furdtran's Gastrointestinal and Liver Disease*, edited by Feldman M, Scharschmidt BF, and Sleisenger MIT, and Saunders WB, 78: 1310, 1993. 47

[224] Rutkowski JM and Swartz MA. A driving force for change: interstitial flow as a morphoregulator. *Trends Cell Biol* 17: 44–50, 2007. DOI: 10.1016/j.tcb.2006.11.007 26, 27, 28, 29, 30, 31

[225] Rutkowski JM, Moya M, Johannes J, Goldman J, and Swartz MA. Secondary lymphedema iin the mouse tail: Lymphatic hyperplasia, VEGF-C upregulation, and the protective role of MMP-9. *Microvasc Res* 72: 161–171, 2006. DOI: 10.1016/j.mvr.2006.05.009 29

[226] Saasoun S and Papadopoulous MC. Aquaporin-4 in brain and spinal cord oedema. *Neuroscience* Aug 12 [Epub ahead of print], 2009. DOI: 10.1016/j.neuroscience.2009.08.019 17

[227] Sabin FR. On the origin of the lymphatic system from the veins, and the development of the lymph hearts and thoracic duct in the pig. *Am J Anat* 1: 367–389, 1902. DOI: 10.1002/aja.1000010310 34, 42

[228] Saetzlera RK, Jalloa J, Lehre HA, Philipsa CM, Vastharea U, Arfors KE, and Tuma RF. Intravital Fluorescence Microscopy: Impact of Light-induced Phototoxicity on Adhesion of Fluorescently Labeled Leukocytes. *J Histochem Cytochem*, 45: 505–514, 1997. 9

[229] Sano Y, Shimizu F, Nakayama H, Abe M, Maeda T, Ohtsuki S, Terasaki T, Obinata M, Ueda M, Takahashi R, and Kanda T. Endothelial cells constituting blood-nerve barrier have highly specialized characteristics as barrier-forming cells. *Cell Struct Funct* 32: 139–147, 2007. DOI: 10.1247/csf.07015 17

[230] Sarelius IH, Kuebel JM, Wang J-J, and Huxley VH. Macromolecule permeability of in situ and excised rodent skeletal muscle arterioles and venules. *Am J Physiol* 209: H474–81, 2006. DOI: 10.1152/ajpheart.00655.2005 11

[231] Scallan JP and Huxley VH. In vivo determination of collecting lymphatic permeability to albumin: a role for lymphatics in exchange. *J Physiol (London)*, 588: 243–254, 2010. 11, 33, 38, 39, 41

[232] Schmid-Schonbein GW. Microlymphatics and lymph flow. *Physiol Rev* 70: 987–1028, 1990. 33, 34, 35

[233] Selman M, Thannickal VJ, Pardo A, Zisman DA, Martinez FJ, and Lynch JP 3rd. Idiopathic pulmonary fibrosis: pathogenesis and therapeutic approaches. *Drugs* 64: 405–430, 2004. DOI: 10.2165/00003495-200464040-00005 29

[234] Semba T, Mizonishi T, Ikeda Y, and Nagao Y. Influence of intestinal inhibitory reflex on mesenteric blood flow through an intestinal segment of the dog. *Jpn J Physiol* 27: 439–450, 1977. 55

[235] Semino CE, Kamm RD, and Lauffenburger DA. Autocrine EGF receptor activation mediates endothelial cell migration and vascular morphogenesis induced by VEGF under interstitial flow. *Exp Cell Res* 312: 289–298, 2006. DOI: 10.1016/j.yexcr.2005.10.029 28

[236] Shields JD, Fluery ME, Yong C, Tomei AA, Randolph GJ, and Swartz MA. Autologous chemotaxis as a mechanism of tumor cell homing to lymphatics via interstitial flow and autocrine CCR7 signaling. *Cancer Cell* 11: 526–538, 2007. DOI: 10.1016/j.ccr.2007.04.020 31

[237] Shimizu F, Sano Y, Maeda T, Abe MA, Nakayama H, Takahashi R, Ueda M, Ohtsuki S, Terasaki T, Obinata M, and Kanda T. Peripheral nerve pericytes originating from the blood-nerve barrier expresses tight junctional molecules and transporters as barrier-forming cells. *J Cell Physiol* 217: 388–399, 2008. DOI: 10.1002/jcp.21508 17

[238] Shirley HH Jr, Wolfram CG, Wasserman K, and Mayerson HS. Capillary permeability to macromolecules: stretched pore phenomenon. *Am J Physiol* 190: 189–193, 1957. 56

[239] Skobe M, Hamberg LM, Hawighorst T, Shirner M, Wolf GL, Alitalo K, and Detmar M. Concurrent induction of lymphangiogenesis, angiogenesis, and macrophage recruitment by vascular endothelial growth factor-C in melanoma. *Am J Pathol* 159: 893–903, 2001. 31

[240] Slaaf DW, Reneman RS, and Widerehielm CA. Pressure regulation in muscle of unanesthetized bats. *Microvasc Res* 33: 315–326, 1987. DOI: 10.1016/0026-2862(87)90026-4 55

[241] Sonsino J, Gong H, Wu P, and Freddo TF. Co-localization of junction-associated proteins of the human blood–aqueous barrier: occludin, ZO-1 and F-actin. *Exp Eye Res* 74: 123–129, 2002. DOI: 10.1006/exer.2001.1100 17

[242] Squire JM, Chew M, Nneji G, Neal C, Barry J, and Michel C. Quasi-periodic substructure in the microvessel endothelial glycocalyx: a possible explanation for molecular filtering? *J Struct Biol* 136: 239–255, 2001. DOI: 10.1006/jsbi.2002.4441 15

[243] Srinivasan RS, Dillard ME, Lagutin OV, Lin FJ, Tsai S, Tsai MJ, Samokhvalov IM, and Oliver G. Lineage tracing demonstrates the venous origin of the mammalian lymphatic vasculature. *Genes Dev* 21: 2422–2432, 2007. DOI: 10.1101/gad.1588407 34, 38, 41, 42

[244] Stacker SA, Caesar C, Baldwin ME, Thornton GE, Williams RA, Prevo R, Jackson DG, Nishikawa S, Kubo H, and Achen MG. VEGF-D promotes the metastatic spread of tumor cells via the lymphatics. *Nat Med* 7: 186–191, 2001. DOI: 10.1038/84635 44

[245] Stamatovic SM, Keep RF, and Andjelkovic AV. Brain endothelial cell-cell junctions: how to "open" the blood brain barrier. *Curr Neuropharmacol* 6: 179–192, 2008. DOI: 10.2174/157015908785777210 17

[246] Starling EH. On the absorption of fluids from the connective tissue spaces. *J Physiol (Lond)* 19: 312–326, 1896. 2

[247] Staub NC and Taylor AE. *Edema*, New York, Raven Press, 1984. 47

[248] Stratman AN, Saunders WB, Sacharidou A, Koh W, Fisher KE, Zawieja DC, Davis MJ, and Davis GE. Endothelial cell lumen and vascular guidance tunnel formation requires MT1-MMP-dependent proteolysis in 3-deminsional collagen matrices. *Blood* 114: 237–247, 2009. DOI: 10.1182/blood-2008-12-196451 24

[249] Stratman AN, Malotte KM, Mahan RD, Davis MJ, and Davis GE. Pericyte recruitment during vasculogenic tube assembly stimulates endothelial basement membrane matrix formation. *Blood*, 2009. [Epub ahead of print]. DOI: 10.1182/blood-2009-05-222364 24, 48

[250] Swartz MA. The physiology of the lymphatic system. *Adv Drug Deliv Rev* 50: 3–20, 2001. DOI: 10.1016/S0169-409X(01)00150-8 34, 42

[251] Swartz MA and Skobe M. Lymphatic function, lymphangiogenesis, and cancer metastasis. *Microsc Res Tech* 55: 92–99, 2001. DOI: 10.1002/jemt.1160 26, 44

[252] Swartz MA. The role of interstitial stress in lymphatic function and lymphangiogenesis. *Ann NY Acad Sci* 979: 197–210, 2002. DOI: 10.1111/j.1749-6632.2002.tb04880.x 26

[253] Swartz MA and Fleury ME. Interstitial Flow and Its Effects in Soft Tissues. *Annu Rev Biomed Eng* 9:229–256, 2007. DOI: 10.1146/annurev.bioeng.9.060906.151850 26, 30

[254] Swartz MA, Hubbell JA, and Reddy ST. Lymphatic drainage function and its immunological implications: From dendritic cell homing to vaccine design. *Seminars in Immunology 20*: 147–156, 2008. DOI: 10.1016/j.smim.2007.11.007 31

[255] Tada S and Tarbell JM. Fenestral pore size in the internal elastic lamina affects transmural flow distribution in the artery wall. *Ann Biomed Eng* 29: 456–466, 2001. DOI: 10.1114/1.1376410 18, 20

[256] Tada S and Tarbell JM. Flow through internal elastic lamina affects shear stress on smooth muscle cells (3D simulations*). Am J Physiol* 282: H576-H584, 2002. DOI: 10.1152/ajpheart.00751.2001 18

[257] Tada S and Tarbell JM. Interstitial flow through the internal elastic lamina affects shear stress on arterial smooth muscle cells. *Am J Physiol* 278: H1589-H1597, 2000. 18

[258] Takahashi T, Shibata M, and Kamiya A. Mechanism of macromolecule concentration in collecting lymphatics in rat mesentery. *Microvasc Res* 54: 193–205, 1997. DOI: 10.1006/mvre.1997.2043 38

[259] Tarbell JM and Ebong EE. The endothelial glycocalyx: a mechano-sensor and –transducer. *Sci Signal* 1: 8, 2008. 15

[260] Tarbell JM, Demaio L, and Zaw MM. Effect of pressure on hydraulic conductivity of endothelial monolayers: role of endothelial cleft shear stress. *J Appl Physiol* 87: 261–268, 1999. 19, 20, 37

[261] Taylor A and Gibson H. Concentrating ability of lymphatic vessels. *Lymphology* 8: 43–49, 1975. 38

[262] Taylor AE, Gaar KA, and Gibson H. Effect of tissue pressure on lymph flow. *Biophys J* 10: 45A, 1970. 36

[263] Taylor AE, Gibson WH, Granger HJ, and Guyton AC. The interaction between intracapillary and tissue forces in the overall regulation of interstitial fluid volume. *Lymphology* 6: 192–208, 1973. 36

[264] Thannickal VJ, Toews GB, White ES, Lynch JP 3rd, and Martinez FJ. Mechanisms of pulmonary fibrosis. *Annu Rev Med* 55: 395–417, 2004. DOI: 10.1146/annurev.med.55.091902.103810 29

[265] Tomei AA, Siegert S, Britschgi MR, Luther SA, and Swartz MA. Fluid Flow Regulates Stromal Cell Organization and CCL21 Expression in a Tissue-Engineered Lymph Node Microenvironment. *J Immunol* 183: 4273–4283, 2009. DOI: 10.4049/jimmunol.0900835 27, 29, 31

[266] Trzewik J, Mallipattu SK, Artmann GM, Delano FA, and Schmid-Schonbein GW. Evidence for a second valve system in lymphatics: endothelial microvalves. *FASEB J* 15: 1711–1717, 2001. DOI: 10.1096/fj.01-0067com 35

[267] Vainionpaa N, Butzow R, Hukkanen M, Jackson DG, Pihlajaniemi T, Sakai LY, and Virtanen I. Basement membrane protein distribution in LYVE-1-immunoreactive lymphatic vessels of normal tissues and ovarian carcinomas. *Cell Tissue Res* 328: 317–328, 2007. DOI: 10.1007/s00441-006-0366-2 33, 42

[268] Van Helden DF, von der Weid PY, and Crowe MJ. Electrophysiology of lymphatic smooth muscle. *London: Portland Press*, 1995. 42

[269] Vollmar B and Menger MD. The Hepatic Microcirculation: Mechanistic Contributions and Therapeutic Targets in Liver Injury and Repair. *Physiol Rev* 89: 1269–1339, 2009. DOI: 10.1152/physrev.00027.2008 18

[270] von der Weid PY. Review article: lymphatic vessel pumping and inflammation–the role of spontaneous constrictions and underlying electrical pacemaker potentials. *Aliment Pharmacol Ther* 15: 1115–1129, 2001. DOI: 10.1046/j.1365-2036.2001.01037.x 41, 42

[271] von der Weid PY and Zawieja DC. Lymphatic smooth muscle: the motor unit of lymph drainage. *Int J Biochem Cell Biol* 36: 1147–1153, 2004. DOI: 10.1016/j.biocel.2003.12.008 33, 40

[272] von der Weid PY, Rahman M, Imtiaz MS, and van Helden DF. Spontaneous transient depolarizations in lymphatic vessels of the guinea pig mesentery: pharmacology and implication for spontaneous contractility. *Am J Physiol* 295: H1989-H2000, 2008. DOI: 10.1152/ajpheart.00007.2008 41

[273] Wang S and Tarbell JM. Effect of Fluid Flow on Smooth Muscle Cells in a 3-Dimensional Collagen Gel Model. *Arterioscler Thromb Vasc Biol* 20: 2220–2225, 2000. 18

[274] Wang S, Voisin MB, Larbi KY, Dangerfield J, Scheiermann C, Tran M, Maxwell PH, Sorokin L, and Nourshargh S. Venular basement membranes contain specific matrix protein low expression regions that act as exit points for emigrating neutrophils. *J Exp Med* 203: 1519–1532, 2006. DOI: 10.1084/jem.20051210 24

[275] Weinbaum S, Tarbell JM, and Damiano ER. The Structure and Function of the Endothelial Glycocalyx Layer. *Annu Rev Biomed Eng* 9: 121–167, 2007. DOI: 10.1146/annurev.bioeng.9.060906.151959 16

[276] Weiss N, Miller F, Cazaubon S, and Couraud PO. The blood-brain barrier in brain homeostasis and neurological diseases. *Biochim Biophys Acta* 1788: 842–857, 2009. DOI: 10.1016/j.bbamem.2008.10.022 17

[277] Wigle JT, Harvey N, Detmar M, Lagutina I, Grosveld G, Gunn MD, Jackson DG, and Oliver G. An essential role for Prox1 in the induction of the lymphatic endothelial cell phenotype. *Embo J* 21: 1505–1513, 2002. DOI: 10.1093/emboj/21.7.1505 42

[278] Wiig H, Reed RK, and Aukland K. Measurement of interstitial fluid pressure in dogs: evaluation of methods. *Am J Physiol* 253: H283–290, 1987. 35

[279] Wiig H, Rubin K, and Reed RK. New and active role of the interstitium in control of interstitial fluid pressure: potential therapeutic consequences. *Acta Anaethesiol Scand* 47: 111–121, 2003. DOI: 10.1034/j.1399-6576.2003.00050.x 3, 25, 27, 47

[280] Wiig H, Gyenge C, Iversen PO, Gullberg D, and Tenstad O. The role of the extracellular matrix in tissue distribution of macromoleculates in normal and pathological tissues: patential therapeutic consequences. *Microcirculation* 15: 283–296, 2008. DOI: 10.1080/10739680701671105 23, 54

[281] Williams DA. A shear stress component to the modulation of capillary hydraulic conductivity (Lp). *Microcirculation* 3: 229–232, 1996. DOI: 10.3109/10739689609148293 20

[282] Williams DA. Intact capillaries sensitive to rate, magnitude, and pattern of shear stress stimuli as assessed by hydraulic conductivity (Lp). *Microvasc Res* 66: 147–158, 2003. DOI: 10.1016/S0026-2862(03)00038-4 20

[283]   Wong CH and Cheng CY. The blood-testis barrier: its biology, regulation, and physiological role in spermatogenesis, *Curr Top Dev Biol* 71: 263–295, 2005. DOI: 10.1016/S0070-2153(05)71008-5 16

[284]   Yuan L, Moyon D, Pardanaud L, Breant C, Karkkainen MJ, Alitalo K, and Eichmann A. Abnormal lymphatic vessel development in neuropilin 2 mutant mice. *Development* 129: 4797–4806, 2002. 43

[285]   Zawieja DC and Barber BJ. Lymph protein concentration in initial and collecting lymphatics of the rat. *Am J Physiol* 252: G602–606, 1987. 38

[286]   Zawieja DC. Contractile physiology of lymphatics. *Lymphat Res Biol.* 7: 87–96, 2009. DOI: 10.1089/lrb.2009.0007 40, 41, 42

[287]   Zaweija DC, von der Weid P-Y, and Gashev AA. Microlymphatic biology. In: *Handbook of Physiology: Microcirculation*, edited by Tuma RF, Duran WN, and Ley K, Chap 5, pp 125–158, Amsterdam, Elsevier, 2008. 40, 41, 42, 45

[288]   Zhang RZ, Gashev AA, Zawieja DC, and Davis MJ. Length-tension relationships of small arteries, veins, and lymphatics from the rat mesenteric microcirculation. *Am J Physiol* 292: H1943–1952, 2007. DOI: 10.1152/ajpheart.01000.2005 40

[289]   Zhang X, Adamson RH, Curry FE, and Weinbaum S. Transient regulation of transport by pericytes in venular microvessels via trapped microdomains. *PNAS* 105: 1374–1379, 2008. DOI: 10.1073/pnas.0710986105 16

[290]   Zweifach BW and Prather JW. Micromanipulation of pressure in terminal lymphatics in the mesentery. *Am J Physiol* 228: 1326–1335, 1975. 41